International Series on Computer Entertainment and Media Technology

Series Editor

Newton Lee
Tujunga, California, USA

The International Series on Computer Entertainment and Media Technology presents forward-looking ideas, cutting-edge research, and in-depth case studies across a wide spectrum of entertainment and media technology. The series covers a range of content from professional to academic. Entertainment Technology includes computer games, electronic toys, scenery fabrication, theatrical property, costume, lighting, sound, video, music, show control, animation, animatronics, interactive environments, computer simulation, visual effects, augmented reality, and virtual reality. Media Technology includes art media, print media, digital media, electronic media, big data, asset management, signal processing, data recording, data storage, data transmission, media psychology, wearable devices, robotics, and physical computing.

More information about this series at http://www.springer.com/series/13820

Patrick C. K. Hung

Editor

Mobile Services
for Toy Computing

Springer

Editor
Patrick C. K. Hung
Fac. Business & Information Technology
University of Ontario Institute of
Technology (UOIT)
Oshawa
Ontario
Canada

ISSN 2364-947X   ISSN 2364-9488 (electronic)
International Series on Computer Entertainment and Media Technology
ISBN 978-3-319-35375-3   ISBN 978-3-319-21323-1 (eBook)
DOI 10.1007/978-3-319-21323-1

Printed on acid-free paper

Springer International Publishing AG Switzerland is part of Springer Science+Business Media
(www.springer.com)

Contents

Introduction to Toy Computing 1
Laura Rafferty and Patrick C. K. Hung

Toy Computing Background 9
Laura Rafferty, Brad Kroese and Patrick C. K. Hung

Mobile Computing Toys: Marketing Challenges and Implications 39
Tirtha Dhar and Terry Wu

**Toys and Mobile Applications: Current Trends and Related
Privacy Issues** 51
Gary Ng, Michael Chow and André de Lima Salgado

**Emerging Human-Toy Interaction Techniques with Augmented
and Mixed Reality** 77
Jeff K. T. Tang and Jordan Tewell

Advanced Sound Integration for Toy-Based Computing 107
Bill Kapralos, Kamen Kanev and Michael Jenkin

Augmented Reality for Mobile Service of Film-Induced Tourism App 129
Wei-Feng Tung

Privacy Requirements in Toy Computing 141
Laura Rafferty, Marcelo Fantinato and Patrick C. K. Hung

**Case Study: Approaching the Learning of Kanji Through
Augmented Toys in Japan** 175
Kamen Kanev, Itaru Oido, Patrick C. K. Hung, Bill Kapralos
and Michael Jenkin

Contributors

Michael Chow Faculty of Business and Information Technology, University of Ontario Institute of Technology, Oshawa, Canada

Tirtha Dhar Faculty of Business and IT, University of Ontario Institute of Technology, Oshawa, Canada

Marcelo Fantinato School of Arts, Sciences and Humanities, University of São Paulo, São Paulo, Brazil

Patrick C. K. Hung Faculty of Business and IT, University of Ontario Institute of Technology, Oshawa, Canada

Michael Jenkin Graduate School of Informatics, Shizuoka University, Hamamatsu, Japan

Electrical Engineering and Computer Science, York University, Toronto, Canada

Lassonde School of Engineering, York University, Toronto, Canada

Kamen Kanev Faculty of Business and Information Technology, University of Ontario Institute of Technology, Oshawa, Canada

Graduate School of Informatics, Shizuoka University, Hamamatsu, Japan

Bill Kapralos Faculty of Business and Information Technology, University of Ontario Institute of Technology, Oshawa, Canada

Graduate School of Informatics, Shizuoka University, Hamamatsu, Japan

Brad Kroese Faculty of Business and IT, University of Ontario Institute of Technology, Oshawa, Canada

Gary Ng Faculty of Business and Information Technology, University of Ontario Institute of Technology, Oshawa, Canada

Itaru Oido Graduate School of Informatics, Shizuoka University, Hamamatsu, Japan

Laura Rafferty Faculty of Business and IT, University of Ontario Institute of Technology, Oshawa, Canada

André de Lima Salgado Computer Science Department, ICMC-University of São Paulo, São Paulo, Brazil

Jeff K. T. Tang School of Computing and Information Sciences, Caritas Institute of Higher Education, Hong Kong

Jordan Tewell School of Mathematics, Computer Science & Engineering, City University London, United Kingdom

Wei-Feng Tung Department of Information Management, New Taipei City, Taiwan

Terry Wu Faculty of Business and IT, University of Ontario Institute of Technology, Oshawa, Canada

Introduction to Toy Computing

Laura Rafferty and Patrick C. K. Hung

Abstract Toys have evolved over the ages from simple dolls and cars to electronic toys, and more recently have been integrating mobile technologies. Toy computing incorporates the physical component of a traditional toy combined with networking and sensory capabilities of mobile devices. Current trends from Toy Fair 2014 and 2015 indicate the growing popularity of augmented reality toys and integration with smartphones and mobile apps. This chapter introduces the concept of toy computing and its place in the industry. Further, this chapter also outlines an architectural model for toy computing illustrating the interactions between the child user, the physical toy component, the mobile device and mobile service.

Keywords: Toy Computing · Mobile Services · BYOD

Introduction

Toys have been a part of human existence for thousands of years, across every culture, being uncovered from as far back as ancient Egyptian times. A toy is an item or product intended for learning or play, which can have various benefits to childhood development. Toys can have a variety of purposes including education, leisure, and socialization. As such a substantial part of the human development, toys have continued to maintain a presence in the daily lives of billions of individuals of all ages. While primitive toys included rocks and pinecones, they soon progressed into dolls, stuffed animals and trains. As new ideas continue to develop to reflect the era and culture, it becomes evident that the toy is a product which has evolved along with humankind. It has become a marketable product which has blossomed into a multi-billion dollar industry. Electronic toys have gained popularity, consisting of electronic parts with embedded systems. In the past few decades, electronic toys such as Speak & Spell, Tamagotchi, and Furby had become popular. More recently,

L. Rafferty (✉) · P. C. K. Hung
Faculty of Business and IT, University of Ontario Institute of Technology, Oshawa, Canada
e-mail: Laura.Rafferty@uoit.ca

P. C. K. Hung
e-mail: Patrick.Hung@uoit.ca

© Springer International Publishing Switzerland 2015
P. C. K. Hung (ed.), *Mobile Services for Toy Computing,* International Series on
Computer Entertainment and Media Technology, DOI 10.1007/978-3-319-21323-1_1

1

sensors, and networking capabilities have introduced a variety of new possibilities for the toy industry. Toy companies have embraced modern technologies such as mobile devices into the design of their products, reshaping the concept of toys and education (LaMonica 2012) through mobile applications and augmented reality.

Trends from Toy Fair 2014 indicated that the future of toys is augmented reality (Bradford 2014; Smith 2014), which uses technology to superimpose virtual world on top of reality. Augmented reality has also been noted to have a significant presence in the toy industry dating back to 2012 (Greenwald 2012). These trends continued into Toy Fair 2015 with RCs that interact with smartphones, apps that allow children to play with toys in new and different ways, augmented reality and wearables (Toy Industry Association 2015). Electronic toys have evolved to become more interactive and personalized to the individual user's preferences and environment by providing services which react to sensing technologies, and can create an augmented reality experience. The NPD's 2013 review of the Global Toy Market (The NPD Group, Inc. 2014) identified the toy category of *youth electronics* as the most prominent subcategory of toys in terms of growth in the U.S. and Europe. This was also a trend identified at Toy Fair 2015 (Toy Industry Association 2015). Further, Euromonitor International's research from 2015 (Tansel 2015) indicates that phenomenal growth in smartphones and tablets stimulate digital gaming, while electronic toys such as children's tablets and cross-platform toys such as amiibo have been a growing trend. In the U.S., the toy industry generates approximately $22 billion in annual retail toy sales, while the total economic impact of the toy industry in the U.S. is over $75 billion as of January 2015 (Toy Industry Association Inc. 2012). Many countries have safety standards and regulations limiting the types of toys that can be sold on the market in order to protect the safety of customers.

Toy computing is a recently developing concept which transcends the traditional toy into a new area of computer research using mobile technologies. A toy in this context can be effectively considered a computing device or peripheral. Toy computing is comprised of two main topics in computer science: physical computing, and mobile services. Physical computing builds upon the traditional idea of the toy by bestowing it with potential embedded systems and sensory devices to create a more reactive and pervasive experience. Mobile devices act as the primary computing device and may also use sensors, while hosting mobile applications and services which complement the physical device. Physical computing combined with mobile services can create an augmented reality environment for toy computing through which users can immerse themselves in the toy computing experience. Augmented reality is the result of overlaying virtual components which react to real-world triggers captured through cameras and/or other sensors. This is achieved through the combination of physical computing with mobile services. The toy may be endowed with sensory and/or networking capabilities, allowing for new opportunities for personalized services based on user preferences and environment. This introduces a Service Oriented Architecture (SOA) approach. One of the key concepts with this type of personalized services is contextual data. The application is able to gather data on the context of the user (e.g. time of day, location, weather) and provide personalized services based on this context data.

In modern times, with the proliferation of personal mobile devices such as smartphones and tablets, many service-oriented distributions have taken on a "bring-your-own-device" (BYOD) model. BYOD is an emerging application distribution model that encourages users to use their own mobile devices to access various online and mobile services. BYOD is being adopted by many traditional consumer electronics products such as electrical appliances and toys. Many studies found out that the traditional services enforcement mechanisms cannot cope with the complex security requirements of the emerging BYOD paradigm because the mobile devices are outside the infrastructure's scope and control (Gartner 2014; Beckett 2014). Moreover, the mobile devices may run third-party services that could, intentionally or unintentionally, violate the safety policy. Most of the related solutions provide a mobile device management service that can block or even reset devices violating the security policy based on blacklist or whitelist approaches in place without advanced behavioral analysis such as motion sensing data in the mobile device. Some toys have become integrated into mobile devices, using apps, sensors and Near-field Communication (NFC). This introduces issues with trust which did not previously exist with traditional electronic toys, which operated on their own trusted platform. With the BYOD model comes additional concerns due to the introduction of an untrusted mobile device intended as the primary trusted system.

Toy Computing Model

Toy computing is a configuration where the physical toy component interacts with a mobile device which connects to one or more mobile services to facilitate gameplay. Figure 1 presents a model of the toy computing environment for the purpose of this book. This model illustrates the interactions between the physical toy component and the mobile device, as well as the interaction between the mobile service

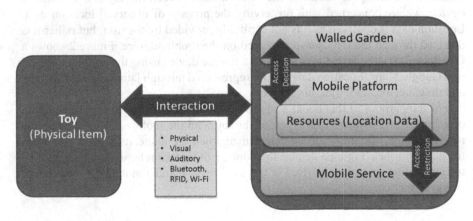

Fig. 1 Toy computing environment

and the mobile platform when the mobile service attempts to access location data resources. In this model we have three entities: the physical toy component, the mobile device (platform), and the mobile service.

The physical toy component is an item much like a traditional toy, and can take the form of anything such as a doll, block, ball, or blaster. The toy interacts with the mobile device through one or more types of interactions. The types of interactions a toy can have with the mobile device include physical interaction by touching a button or screen, visual interaction as detected through a camera on the mobile device, audible interaction as detected through a microphone on the mobile device, or through network interactions such as Bluetooth, RFID or Wi-Fi. The physical toy component may also have sensors which collect and send sensory data to the mobile device.

The mobile component of the model includes the mobile platform, location resources, and the mobile service. The mobile platform facilitates access restrictions between the mobile service and the resources located on the mobile device. Location data is often used by mobile services to provide relevant location-based services to a user. In a toy computing environment, location data can be used to locate other players. Location data is in the form of Global Positioning System (GPS) coordinates, latitude and longitude, to indicate the physical location of the device. Location accuracy can be expressed as fine or coarse, depending on the method it is collected.

This model incorporates the BYOD *Walled Garden* concept as outlined by the Whitehouse (Digital Services Advisory Group and Federal Chief Information Officers Council, United States of America 2012), to contain data or application processing within a secure application on the personal device so that it is segregated from personal data. As depicted in Fig. 1, the mobile service is within a Walled Garden. A Walled Garden in a BYOD environment provides a trusted platform which can make access control decisions based on policies.

Referring to the three types of context data as defined by the World Economic Forum (World Economic Forum 2011), context data can be volunteered, observed, or inferred. During the course of the interactions between the toy and the mobile device, we are concerned with preserving the privacy of observed location data. Location data is observed as it is not explicitly provided by the user, but rather it is detected through the GPS sensors located on the mobile device. Figure 2 shows a GPS location (latitude and longitude) on a mobile device using the popular application Google Maps. Location data will be represented through latitude and longitude, and time. i.e. location=(lat, long, time).

For the purpose of this book, a session is the duration of a game. The duration of a game starts with the interaction of the toy with the mobile device. The access to the location resources will be determined when the game starts, and end when the game finishes. Figure 3 depicts the concept of a session between a user and the toy computing system, comprised of the physical toy, and a mobile device running a mobile service.

Fig. 2 GPS location on a mobile device using Google Maps. (Adapted from play. google.com)

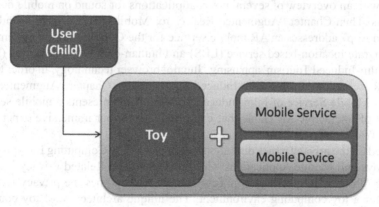

Fig. 3 User and toy computing system

Book Organization

This book is organized into nine main chapters. Chapter "Toy Computing Background" discusses a comprehensive background on toy computing and related topics including physical computing and mobile services technologies. Chapter "Toy Computing Background" includes a discussion of some current toy computing products on the market, as well as a technical background on related technologies, an overview of access control and privacy concepts and a discussion of related works in this research area for Chapter "Augmented Reality for Mobile Service of Film-Induced Tourism App". Then Chapter "Mobile Computing Toys: Marketing Challenges and Implications" presents the marketing issue of toy computing by conducting a brief survey of the traditional toy and trying to identify the key drivers of success and failure of such toys. The paper also addresses a number of tools of marketing research to explain how traditional toy makers can leverage information and mobile technologies to re-establish themselves in a sustainable growth path. As smartphones and mobile applications play a larger role in the everyday lives of their users, toy manufacturers commonly face the challenge of better understanding the modern consumer's needs as well as their concerns. Chapter "Toys and Mobile Applications: Current Trends and Related Privacy Issues" presents the current trends in toys and related privacy issues for Chapter "Augmented Reality for Mobile Service of Film-Induced Tourism App" as well as discusses a case study of toy guns in Brazil.

Chapter "Emerging Human-Toy Interaction Techniques with Augmented and Mixed Reality" gives a review of the emerging technologies, devices and products that promote extensive interactions between toys and their players in Augmented Reality (AR) and Mixed Reality (MR). Next Chapter "Advanced Sound Integration for Toy-Based Computing" provides an overview of sound and spatial sound for use in human-computer interfaces with a particular emphasis on its use in mobile devices and toys. Chapter "Advanced Sound Integration for Toy-Based Computing" concludes with an overview of several novel applications for sound on mobile devices and toys. Then Chapter "Augmented Reality for Mobile Service of Film-Induced Tourism App" addresses an AR mobile service for the Google Android system that can integrate location-based service (LBS) and human-computer interaction (HCI) for a 'Film-Induced Tourism' app using 'Junaio' browser technology. In order to extend AR technology for the Film-Induced Tourism App, Chapter "Augmented Reality for Mobile Service of Film-Induced Tourism App" presents a mobile service concept of interactive AR Game that can provide a further immersive experience App or Web game.

Based on the privacy issues discussed in Chapters "Toy Computing Background" and "Toys and Mobile Applications: Current Trends and Related Privacy Issues", Chapter "Privacy Requirements in Toy Computing" outlines the privacy requirements for a toy computing environment. The unique architecture of toy computing requires consideration of several different factors. Chapter "Privacy Requirements in Toy Computing" investigates the privacy requirements through formal threat modeling techniques to help the reader to get more comfortable with the toy computing architecture and how it maps to privacy threats. In addition, a demo is presented as an interface for parents to configure privacy settings for their children using mobile toy computing apps. Lastly Chapter "Case Study: Approaching the Learning of Kanji through Augmented Toys in Japan" considers the educational perspectives of various toys and toy technologies in Chinese Characters (kanji) learning in Japanese language, and presents related instructional features and learning functionalities. A research initiative led by the authors that integrates traditional toys with novel augmented reality technologies to support the learning of kanji characters, a time consuming and often difficult task, is reported and discussed.

References

Beckett, P. (2014). BYOD – Popular and Problematic. *Network Security, 2014*(9), 7–9.

Bradford, K. T. (2014, February 27). *The Future of Toys Is Augmented Reality*. Retrieved December 2014, from Alley Watch: http://www.alleywatch.com/2014/02/the-future-of-toys-is-augmented-reality/

Digital Services Advisory Group and Federal Chief Information Officers Council, United States of America. (2012, August 23). *Bring Your Own Device*. Retrieved September 2014, from http://www.whitehouse.gov/digitalgov/bring-your-own-device

Gartner. (2014). *Key Challenges in BYOD*. Retrieved September 2014, from http://www.gartner.com/technology/topics/byod.jsp

Greenwald, W. (2012, February 14). *Augmented Reality Takes Center Stage at Toy Fair 2012*. Retrieved 2014 December, from http://www.pcmag.com/article2/0,2817,2400230,00.asp

Health Canada. (2012). *Industry Guide to Health Canada's Safety Requirements for Children's Toys and Related Products*. Health Canada. Retrieved from http://www.hc-sc.gc.ca/cps-spc/pubs/indust/toys-jouets/index-eng.php#f1

LaMonica, M. (2012, April 18). *Mobile Apps Reshape Toys and Learning*. Retrieved December 2014, from http://www.onet.com/news/mobile-apps-reshape-toys-and-learning/

Smith, A. (2014, February 19). *Tiny Sensors are a Game Changer at Toy Fair 2014*. Retrieved 2014, from http://www.huffingtonpost.com/andrea-smith/tiny-sensors-are-a-game-c_b_4814579.html

Tansel, U. (2015). Toys and Games: Global Trends, Developments, and Prospects. Euromonitor International. Retrieved from http://blog.euromonitor.com/2015/01/global-trends-developments-and-prospects-in-toys-and-games.html

The NPD Group, Inc. (2014). *2013 Review of the Global Toy Market*. Toy Industry Association. Retrieved from http://www.toyassociation.org/App_Themes/tia/pdfs/facts/ToyMarkets13.pdf

Toy Industry Association. (2015). *Toy Trends*. (Toy Industry Association) Retrieved 2015, from http://www.toyassociation.org/TIA/Industry_Facts/trends/IndustryFacts/Trends/Trends.aspx

Toy Industry Association Inc. (2012). *Economic Impact of the Toy Industry 2013 Summary Report*. Toy Industry Association Inc. Retrieved from http://www.toyassociation.org/app_themes/tia/pdfs/economicimpact/unitedstates.pdf

World Economic Forum. (2011). *Personal Data: The Emergence of a New Asset Class*. World Economic Forum.

Toy Computing Background

Laura Rafferty, Brad Kroese and Patrick C. K. Hung

Abstract The purpose of this chapter is to provide a background on the fundamental concepts of Toy Computing, including mobile services, physical computing, and augmented reality. It will also present some examples of toy computing products currently on the market. The chapter will also provide a background on security and privacy and relevant research works on these topics in order to provide the necessary foundations for the rest of this book.

Keywords Toy Computing · Mobile Services · Physical Computing · Location Privacy

Section 1. Toy Computing Background

What is Toy Computing?

Mobile devices have become prevalent in many aspects of our daily lives. The reason for this is the portability and flexibility of the devices which can easily support applications developed for a wide range of uses. More recently, another use for mobile devices has been introduced in the area of toys and gaming. Toy companies such as Hasbro, Mattel, and Tech4Kids have released toys that integrate with mobile platforms, providing new capabilities and add-ons to traditional functionality (D'Hooge und Goldstein 2001). These have been referred to as *Augmented* (Hinske & Langheinrich, Managing Augmented Toy Environments–A New Perspective for Smart Space Management 2007), *Interactive* (Luckin et al. 2003), or *Smart Toys* (Plowman und Luckin, February 2004), because they include sensory capabilities to allow them to detect and interact with their environment. Related fields include

L. Rafferty (✉) · B. Kroese · P. C. K. Hung
Faculty of Business and IT, University of Ontario Institute of Technology, Oshawa, Canada
e-mail: Laura.Rafferty@uoit.ca

B. Kroese
e-mail: brad_kroese@hotmail.com

P. C. K. Hung
e-mail: Patrick.Hung@uoit.ca

© Springer International Publishing Switzerland 2015

P. C. K. Hung (ed.), *Mobile Services for Toy Computing,* International Series on Computer Entertainment and Media Technology, DOI 10.1007/978-3-319-21323-1_2

Fig. 1 Toy computing
components

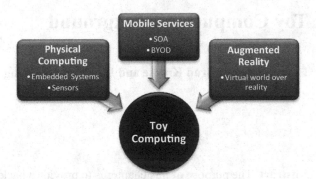

physical computing, mobile services, context and location-based services, and augmented reality. At its most basic level, a toy computing system can be identified as a toy equipped with sensory technology, mobile computing power, and communication capabilities (Hinske & Langheinrich, Managing Augmented Toy Environments–A New Perspective for Smart Space Management 2007). This differs from a traditional electronic toy in how it incorporates a mobile component, whereas traditional electronic toys are isolated to their own proprietary platform. The two basic components that make up a toy computing system include a) the physical component, which is similar to a traditional toy, and b) the mobile component, a smartphone or tablet running an application to provide services to the user/toy.

The physical component of a toy computing system observes almost the same overall characteristics as a traditional toy, with the potential addition of embedded systems, networking capabilities or sensors designed to communicate in some way with the mobile component. This physical component can take the form of any traditional toy, such as a blaster (Tech4Kids 2013), block (ChineseCUBES 2014), or stuffed animal (Woollaston 2014). The physical component may or may not contain embedded systems or networking capabilities; however it must be able to interact in some way with the mobile component. An interaction can be physical, visual, auditory, or through networking such as Bluetooth, RFID or Wi-Fi.

In this configuration, the mobile device takes on the position as the primary computing device of the system. This includes the CPU, memory, sensory input, and output. The mobile component will run an application which operates in collaboration with the physical component to provide services to the user based on their interactions with the physical component. For the purpose of this work, we will be concentrating on *Toy Computing* from a mobile services perspective. There is a multitude of built-in sensory capabilities on mobile devices, which provide a new wave of opportunities for human computer interaction and personalized context-aware services. Depending on the toy, the sensory capabilities may either be located on the physical component, the mobile component, or both. Figure 1 illustrates the relationship between physical computing and mobile services to form a toy computing environment.

Examples of Toy Computing Products

Toy computing is quickly gaining popularity in the toy industry. These toys have a wide variety of categories including toy blasters, language blocks for educational

Fig. 2 Tek Recon "Havoc" blaster with mobile device mount. (Adapted from www. tekrecon.com)

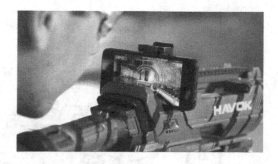

purposes, and methods of communication for children. Below are some examples of popular toy computing products currently on the market.

Tek Recon

Tek Tecon (Tech4Kids 2013) is a line of toy blasters developed by Tech4Kids, marketed to children aged 8 years and up in 2013. While this product features a physical component identical in concept to a traditional toy blaster, the novelty is the ability to integrate with a mobile device. Referring to Fig. 2, the Tek Recon blaster features a mount on top where a smartphone is inserted. A mobile application has been developed by Tech4Kids which operates in collaboration with the physical blaster to augment traditional blaster-based games. The application provides several functionalities including a scope, which uses the smartphone camera to display what is in front of the user with additional features overlaid on top, such as ammunition, score, radio, and a GPS location map of other players. The application has networking functionality to create and join games with friends over a LAN or mobile network. The user is also required to create an account online, where the scores and account information are stored.

Sphero

Another recent toy computing product in the industry is Sphero (Sphero 2014), first introduced in 2011 by Orbotix, which then released subsequent versions, Sphero 2.0 in 2013 and Sphero Ollie in 2014. Referring to Fig. 3, Sphero is a robotic ball which can be controlled and programmed through the user's smartphone or tablet. There are over 30 apps available for Sphero, most of which are games, while others

Fig. 3 Sphero robotic ball. (Adapted from www.think-geek.com)

Fig. 4 ChineseCUBES. (Adapted from www.chinesecubes.com)

are focused on education. This product is marketed not only to children and can be appropriate for any age group. While the physical ball component is a very simple and traditional concept, the capabilities of the toy increase substantially with the inclusion of robotics and a mobile device. The Sphero ball has wireless networking capabilities, an accelerometer and gyroscope, rolls in every direction, and glows different colors. Sphero can be programmed by the user through an app called Sphero Macrolab, which includes a set of predefined macros, and more advanced users can use another app called orbBasic to program in a language based on BASIC.

ChineseCUBES
ChineseCUBES (ChineseCUBES 2014) is a toy computing product first introduced in 2011 which combines augmented reality technology with physical blocks to help the user to learn Chinese characters. Referring to Fig. 4, the AR markers on the cubes are arranged in a certain order and detected by the webcam to create an interactive audio/visual experience with the software on the computer or mobile device. The software includes multiple features such as interactive stories, lessons and videos. The compute or mobile device does all of the sensing and processing in this scenario, and the physical cube components are entirely traditional.

Toy Mail
Toy Mail (Toymail Co. LLC 2014) is a toy introduced in 2013 which can connect to the user's home WiFi network and interact with the free Toy Mail mobile app. Once the app is installed on a mobile device, the user can record a message which will be sent to the toy. When a message is received, the toy (as shown in Fig. 5) will make a snort, wheeze, or whine sound to let the user know that they have received a message, which can then be played and replied to.

Other
Toy computing has been also been developed for a wide range of purposes such as language learning (Lee und Doh 2013), early childhood education, and for children

Fig. 5 Toy mail character, "Snort." (Adapted from www.toymail.co)

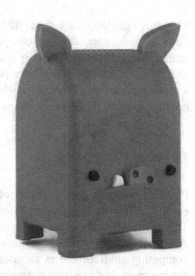

with ADHD and autism. For example, Auti is a socially assistive robotic toy which encourages physical and verbal interactions in children with autism (Andreae et al. 2014). Educational toys such as roBlocks and SmartTile encourage children to learn about robotics and programming while they play (Gross und Eisenberg 2007). There has also been research on monitoring children's developmental progress using augmented toys and activity recognition (Westeyn et al. 2012).

Design Guidelines for Toy Computing

Hinske et al. (Hinske et al. Towards Guidelines for Designing Augmented Toy Environments 2008) provide a Summary of Design Guidelines for Integrating Pervasive Computing Technology into (Traditional) Toys:

1. The technological enhancement must have an added value.
2. Specify what actions/tasks are to be supported.
3. Let the focus remain on the toy and the interaction itself, not the technology.
4. Integrate the technology in such a way that it is unobtrusive, if not completely invisible.
5. Toys should be still usable (in the "traditional" way) even if the technology is switched off or not working.
6. Tightly intertwine design and implementation
7. The technology should be reliable, durable, and safe.
8. Offer immediate and continuous feedback.
9. The added technology should support the high dynamics of play environments.
10. Employ an iterative development process, including rapid prototyping and testing.

The above guidelines reinforce that the integration of pervasive computing technology (i.e. in the context of toy computing) should provide added value and seamless integration with the physical toy component. Further, the technology should be reliable, durable, and safe. From the perspective of privacy, this introduces a need for a privacy preserving framework which protects the child from privacy threats while not taking away from the play experience by introducing obtrusive policies.

Physical Computing

Physical computing is a branch of computing which involves the integration of computing technology into a physical device which interacts with its environment. This is similar to the concept pervasive or ubiquitous computing, in which the computing device establishes itself into the users' daily physical activities. A pervasive computing environment is an information-enhanced physical space, not a virtual environment that exists to store and run software (Saha 2003); where the design of the system takes a human body as a given, and attempt to design within the limits of its expression (Sheth et al. 2013).

The distinction between physical and pervasive computing is that physical computing has more of a focus on the physical objects involved rather than completely seamless interaction. In toy computing the physical toy component is an active part of the user experience, whereas in pervasive computing there would be little to no physical component and the system works seamlessly with everyday activities. One of the main characteristics of both pervasive and physical computing devices is its ability to perceive context information on the surrounding environment in order to react accordingly (Saha 2003). This perception is done through sensors on the device such as a microphone, camera, or accelerometer. Perception of this context information is fundamental to the device's ability to make timely and context-sensitive decisions.

As mentioned previously, the physical component of the toy computing environment would be the traditional toy itself, which will be complemented with embedded systems or sensor technology which communicates with the mobile application. In this system, personalized services are provided to the user based on context data collected and inferred through sensors and other environment data. With the pervasiveness of modern mobile devices, vast amounts of information can be collected and inferred about the user and their environment. Physical computing often involves a networked environment, which introduces privacy and security issues, particularly related to the context information the devices are processing. While the toy is the physical component in this system, the mobile device is what provides computing functionality and sensory perception, as described in the next section.

Physical computing introduces physical objects as interface components. The examples in the previous section demonstrate this with a toy blaster, ball, and cubes, which are all used as an interface similar to a traditional toy, but with enhanced interactive capabilities. As seen in Fig. 6, the Sphero robotic ball acts as the physical interface component in a physical, toy computing environment.

Fig. 6 Sphero robotic ball
as a physical toy compo-
nent. (Adapted from www.
gosphero.com)

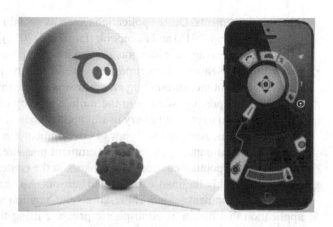

Sensors

Modern mobile devices are created with a variety of sensory capabilities. In a toy computing environment, developers may embed sensors into the physical toy component, or take advantage of sensors already built into the mobile device. Through these sensors, motion and other data can be detected in a number of ways. Sensors can be categorized into three different types: motion sensors, position sensors, and environment sensors (Google n.d.). Below is an analysis on some of the different types of data that can be gathered from these sensors.

- **Motion Sensors**: Motions sensors capture the physical motions of a device. Mobile devices can include a number of sensors for measuring motion including an accelerometer, gyroscope, magnetometer, barometer, gravity, linear acceleration, and rotation vector. Motion is commonly represented through 6- or 9-axis sensor system (3-axis magnetometer, 3-axis gyroscope, 3-axis accelerometer). These types of sensors are commonly used for a variety of mobile applications such as games, as a way for the user to interact with the application (e.g. angling the device left or right to turn the character in a game). They have also been commonly used in fitness applications for tracking steps and calories lost during a walk, run, or jog. A popular example of this is *Zombies, Run!* (Six to Start n.d.), a mobile game application which takes motion sensor input while a user is running or jogging. The application provides missions for the user to complete by meeting certain fitness goals which correlate with the storyline.
- **Position Sensors**: Position sensors are also very popular in mobile systems. Some examples of these types of sensors include geomagnetic field sensor, proximity sensor, and GPS. Position sensors, particularly proximity and GPS, are very useful for mobile and toy computing due to the portability of mobile devices. Many applications use location-based services which use position sensors on the device to provide recommendations relevant to the location of the user. Some examples of this include Yelp (Yelp 2015), UrbanSpoon (zomato 2015), which allow users to read and post reviews of nearby restaurants and

other establishments. Other applications such as social media applications, Instagram (Instagram 2015) and Facebook (Facebook 2015), use position sensors to allow users to geotag their location along with their posts.

- **Environment Sensors**: Sometimes it is useful for an application to be able to detect data about the surrounding environment. While this is not used as widely as motion and position sensors in the mobile and toy computing environment, these sensors do have a lot of very useful applications in agriculture, health care, security systems, aeronautics. Types of environment sensors include sensors for relative ambient humidity, luminance, ambient pressure, and ambient temperature. The most popular environment sensors in the context of a mobile environment are probably luminance and sound sensors. An example of an application that uses environment sensors is PressureNet (Cumulonimbus 2015), an Android application that measures atmospheric pressure using the atmospheric sensors built into most Android phones. Most smartphones use luminance sensors to adjust screen brightness based on lighting conditions.

Wireless Communication Technologies

While physical computing environment collects environment data through sensors, the data collected often needs to be communicated to a service provider or other devices over a wireless network. The service provider may be located on the user's mobile device, or another device on the local or wide-area network. Possible types of wireless communication technologies used in a mobile toy computing environment include: RFID, NFC, Bluetooth, WiFi, GSM, and UMTS/3GSM.

Context Data

Data observed and collected through sensors gather context on the user and their environment. Context is defined succinctly by Dey and Abowd (Dey und Abowd 1999) as "any information that can be used to characterize the situation of an entity." Schilit et al. (Schilit et al. 1994) defined context as location, identities of nearby people and objects, and changes to those objects. Zimmermann et al. further categorized the elements for describing context information into five categories: individuality, activity, location, time, and relations. Individuality is personal information about a user, activity is data regarding physical activity, location is the GPS location, time is discrete time, and relations are inferences between two or more pieces of context data. In a context-aware system, services are provided to the user based on what is relevant to their context. Recent advances in mobile technology open up great opportunity for the collection and processing of context data in valuable ways. There are many types of private context data that can be collected via a mobile application. The collection of this data allows applications to adapt to the user's environment and personalize services accordingly.

Types of Context Data

Mobile devices can capture a user's physical activity state (e.g. walking, standing, running, etc.) and store personalized information (e.g. location, activity patterns, etc.). This data is referred to as context data; data that is collected on the user and their environment. This data can be collected from sensors, provided explicitly by the user, or observed, such as the time of an event. Personal data can come in many forms including browsing history, friends list, and location information. Some other examples of relevant context information include (Schmidt 2005): Verbal context, roles of communication partners, goals of the communication/individuals, local environment, social environment (who is there), and physical and chemical environment. Information can be volunteered (e.g. profile data provided directly by the user) or observed (e.g. location data detected from GPS). Often, private information may seem trivial and not perceived as very sensitive to the user, while in practice it can actually reveal a large amount of personal information about them. The World Economic Forum (World Economic Forum 2011) defines three types of context data, as categorized by the way it is collected: volunteered, observed, and inferred:

- **Volunteered Data**: data that is explicitly provided by the user. This can include personal profile information or preference settings.
- **Observed Data**: data not directly given by the user, but is detected by the device/ application often through a sensor. Some examples of observed data include GPS location and time.
- **Inferred Data**: data deduced based on analysis of a combination of volunteered and/or observed data (e.g. where a user is likely to be going based on typical behavior). A lot can be interpreted on a user and their environment through inferences based on collected data. There is great value on this inferred data that would not be explicitly provided by the user.

Volunteered and observed data can be analyzed to infer significant amounts of personal information about the user. For example, forecasting trip destinations based on data from driving habits (Dewri et al. 2013). Collected data is the basis for many valuable context-aware services, which provide custom content or services to the user based on what is most likely to be useful to them.

Privacy Concerns

With all of this in mind, privacy is a growing concern among many users of mobile devices. While many users appreciate the value of targeted services, they still express concern over how their data is collected and managed without their knowledge. Cherubini et al. (Cherubini et al. 2011) identify privacy as a barrier to the adoption of mobile phone context services. 70 % of consumers say it is important to know exactly what personal information is being collected and shared (MEF 2013), while 92 % of users expressed concern about applications collecting personal information without their consent (Futuresight 2011). Mobile applications have adapted countless services to better analyze context data and provide custom services that will bring the most value to a user based on what they are most likely to need.

While allowing context data to be collected for services can prove to be of great benefit to users, there is an ongoing tradeoff between utility and privacy (Chakraborty et al. 2013b). In this physical mobile and pervasive environment, the timely delivery of services is fundamental. The amount of information collected often results in a tradeoff required between disclosing sensitive data and receiving context-aware services. In order to provide the most relevant services to the user, more personal and context information must be collected, which raises concerns of privacy. For example, a service can send special promotions and coupons to a user depending on what is most relevant to them. In order to provide the most relevant promotions, the service will need to collect certain context data such as their location, and also potential profile information such as age and gender to help to determine what their interests may be based on demographic. To gain even more context of the user, the application may collect and retain historical data on the user such as previous movement patterns, to determine where they are likely to be at certain times, if they are travelling, or previous interactions with the application such as which promotions they had previously been interested in. In this example, it is clear that the more information is collected on the user, the more relevant services can be provided to them. However, the user may not be comfortable with the level of data that is collected and inferred on them. An application knowing where you are and what you are likely to be doing at any given time is likely to raise concern with users.

For this reason, context data is at the core of privacy concerns with many mobile applications. Privacy goals must be defined to ensure private data is managed responsibly. Further, detailed analysis is required to ensure that the user's sensitive behavior cannot be inferred based on collected data. There have been many solutions which aim to preserve the privacy of sensitive context data, as will be described further below. There are countless types of data that can be collected from a mobile device that must be considered when evaluating the scope of privacy. This is true of collected sensory data as described above, and also from within other applications, sensitive data can be collected such as a user's profile information, contact list, or calendar. All of this information can be collected and analyzed to determine context information about the user.

Location

Location data can be defined as data representing where a user is physically located. Location is one of the most prominent types of data for context-based services, existing as a key parameter to define context (Schmidt et al. 1999). A user's location, combined with other context and historical data, can be used to infer an extensive amount of information including actions, speed, direction, and movement patterns. Location data can be collected from the device through GPS, WiFi, or mobile network satellite. It can also be inferred from other information such as IP address, although this can be inaccurate (e.g. in the case of a proxy).

Location is defined succinctly by Merriam-Webster (Merriam-Webster n.d.) as "a place or position." This definition has been extended by the National Geographic Encyclopedia (National Geographic n.d.) to establish three different types of representing location: absolute location, relative location, and type of location as follows:

- **Absolute Location**—the location expressed in a range or exact GPS coordinates of latitude and longitude. The absolute location can be expressed as coarse or fine; for example, an entire country, city, block, or exact coordinates.
- **Relative Location**—the location relative to another entity as a reference point; for example, a relative location can be expressed as the distance between User A and User B, or distance between User A and Device C, or User A and location D.
- **Type of location**—the location expressed in an assigned category. Some examples of this could be home, office, street, mall, or restaurant.

Generally, location is represented as a 3-dimensional vector of GPS coordinates (latitude, longitude), and altitude (optional). A location event also includes a timestamp. Android's **GpsLocation** data structure represents the location of the device with the following data fields [1]:

- Size
- Flags
- Latitude
- Longitude
- Altitude
- Speed
- Bearing
- Accuracy
- Timestamp

Different ways of collecting location information can be more accurate than others. For example, there is GPS-based location (fine) or Network-based location (coarse) (Android 2015):

- ACCESS_COARSE_LOCATION (Network-based)—allows and app to access approximate location derived only from network location sources (cell towers and Wi-Fi). This method varies in accuracy from 50 m in urban areas, and several kilometers in rural areas with less cell tower coverage.
- ACCESS_FINE_LOCATION (GPS-based)—allows and app to access precise location from location sources such as cell towers and WiFi, and also the user's GPS coordinates provided from their device. Accuracy for fine location using GPS is fairly accurate from 2 to 20 m.

While a huge number of mobile applications request access to user location data, this is one of the most sensitive types of context-data. The incredible amount of information that can be gathered from a user based on their location is immense. Whalen et al. (Whalen 2011) discuss some of the current privacy issues in mobile devices mainly focusing on the storing and transmitting of sensitive location based information over extended periods. This research states that a large amount of users do not even know that such information is being stored, and in some cases, still happens even if the user has explicitly restricted such data to be collected. This goes against the privacy principle of having the users consent before collecting this information. Another violation of privacy principles that is discussed in this paper is the amount of data that is being collected is much more plentiful, accurate, and goes on for a lot longer than it needs to. One of the causes of this disconnect is that most of the permissions for this information collection is buried in lengthy policies that users rarely read, and is enabled by default. It is very important to protect this information, having access to such information not only shows where we have been but it can be used to predict where we will be tomorrow, and that introduces a lot more security concerns. Patil et al. (Patil et al. 2012) go into further detail about the widespread usage of location data collection in mobile services and their interaction with social networking services. The paper details an online study on 362 participants to understand the preferences of users of location services. The majority of users expressed that their main incentive for using of these services was for social networking purposes. A number of users in this study (25 %) also indicated that they have regretted sharing their location on at least one occasion.

Location privacy is a huge concern in the mobile and wireless environment. While it can appear trivial, location-based data can infer a lot of sensitive information about a user, including their activities, habits, interests, and personal relationships. Inference attacks are possible, such as knowing when a user will be somewhere based on movement patterns and historical activity. This can potentially put a user at risk. Often, location-aware services do not require knowledge of the exact location, but rather, could provide just as valuable services with an approximation of the user's location (Pandit und Kumar 2012).

These research works identify a great need for location privacy management and enforcement in mobile services. While toy computing has become a recent development in the union of mobile service and toys, research on safety and privacy guidelines for toy computing seems to have been largely overlooked. To the best of our knowledge, there is a gap in the research area of location privacy in the context of toy computing. First, there is no formal model for enforcing privacy or location privacy in particular, for children using such toys.

Section 2. Mobile Services

Mobile devices, such as smartphones, tablets, and e-readers, have become increasingly popular in recent years, successfully integrating themselves into the lives of many users. A recent poll of 5000 people by TIME magazine reveals that 54 % of respondents check their mobile device at least once an hour (TIME and Qualcomm 2012), while another study for GSMA shows that 68 % of participants identified themselves as users of mobile internet/apps, with 38 % of this subset considering themselves to be heavy users (Futuresight 2011). The immense popularity of these devices can be explained by their personal, portable, and pervasive nature. These characteristics create a unique platform for services, particularly those based on context data. Often, mobile devices will have one single primary user. The portability of a mobile device makes it possible for a user to carry it with them wherever they go, making it a highly personal device as well. Mobile devices are also designed to be easy and fast to use, and easily connect to networks, allowing the user to always stay connected to data and services.

Mobile devices use mobile services, which are services accessible through a mobile network. Mobile services, like Web services, use Service Oriented Architecture (SOA) as described in Sect. 2.2.3.1. Mobile services can be context-aware, gathering context information from the mobile device, and providing relevant personalized services based on the context. To gather context information, a context-aware service can either listen for events sent by a context provider, or query the context provider. Gu et al. (Gu et al. 2004) propose a middleware for building context-aware mobile services, using a Service Locating Service to allow entities to locate different context providers. However, this model does not consider privacy preferences of the user.

Mobile Games and Location-Based Services

Location-based services, also known as location-aware mobile services, have become widely popular to provide information such as travel information, shopping, entertainment, and event information. Location-based services have been defined by Duri et al. (Duri et al. 2001) as "services in which the location of a person or an object is used to shape or focus the application or service." Pura (Pura 2005) identifies location as one of the most promising applications of mobile commerce, due to the ability to allow service providers to offer customized services based on context and resulting in increased perceived value and loyalty of customers.

The mobile application industry has observed a widespread adoption of mobile game applications. This has been successful due to factors such as increased mobility and social network integration (Baber und Westmancott 2004). Location-based services have also been used in applications for games. The popular mobile game Angry Birds (Rovio 2015) has a location-based feature which allows users to compete with other based on a leader board associated with their location. MyTown (Booyah 2015) is another mobile game, reminiscent of Monopoly, where users can check in to a physical location, buy and sell properties, and collect rent from other players who check into the same location.

Kaasinen (Kaasinen 2003) conducted a study to investigate user needs for location-aware mobile services:

- Contents: topical up to date information, comprehensive relevant information, interaction (user is moving and can only provide limited interaction to device), push information based on both location and personalization, detailed search options, planning vs. spontaneity.
- Personalization: personal options and contents, user-generated content.

- Seamless service entities: consistency, seamless solutions to support the whole user activity.
- Privacy: the right to locate, use, store, and forward the location. Privacy requirements are based on legislation and social regulation. The paper also identifies P3P as a potential approach to manage user privacy preferences and compare them to the location-aware service's privacy practices.

Bring Your Own Device (BYOD)

Mobile services follow Bring *Your Own Device* (BYOD) architecture, meaning that the user has their own personal mobile device to run the service from. Mobile services thus need to be flexible and consider a variety of different devices. While the term BYOD is typically used to refer to employees bringing personal devices to a work environment, the same general idea is involved in any mobile services scenario. Mobile applications must operate in a controlled environment and must protect data and resources from other untrusted applications that may be running on the device. Further, the introduction of unregulated mobile devices onto a network can result in loss of control, data leaks, and potential network loss (Gartner 2014). BYOD can introduce complications when it comes to investigation in the case of a security breach. This can be made simpler through thorough planning of policies and contracts indicating employee and employer (or in a more general case, user and service provider) rights (Beckett 2014).

A toy computing environment considers several properties of BYOD, although outside of a corporate environment. For the purpose of this work, we will be considering the following BYOD characteristics:

1. The user's mobile device is untrusted.
2. The mobile application is operating on top of this untrusted device.

The objective of BYOD is to isolate business applications from the rest of the system. This means isolation from other applications running on the personal device (Disterer und Kleiner 2013).

Mobile Service Architecture

Figure 7 illustrates a multi-layered model which illustrates the relationship between the conceptual, logical, and language layers of mobile services. This framework has been adapted from the Web services logical model presented by Hung et al. (Hung et al. Towards Standardized Web Services Privacy Technologies 2004). This model is an extension of traditional Service-Oriented Architecture to include layers for privacy-related access control, and also an End-Point Device Profile for mobile devices. Each layer will be discussed in the subsequent sections.

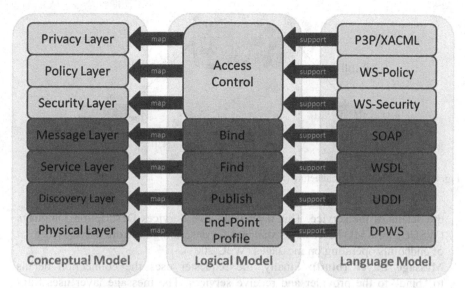

Fig. 7 Mapping between different models and layers. (Adapted from Hung et al. Towards Standardized Web Services Privacy Technologies 2004)

Service Oriented Architecture (SOA)

A theoretical model for Web services has been defined in Service Oriented Architecture (SOA). In our conceptual model, SOA consists of the message layer, service layer, and discovery layer. W3C defines SOA as a form of distributed systems architecture which typically maintains the following six properties (W3C 2004): (1) The architecture is defined in a *logical view*, in terms of what it does. (2) The *message orientation* property expresses how the service is defined in terms of the messages exchanged between provider and requester agents, rather than the internal architecture behind the provider's services. (3) *Description orientation* enforces that a service is described by machine-processable metadata. (4) SOA messages are also *granular* and (5) P*latform neutral*, meaning services tend to use a small number of operations with large and complex messages, which are in a standardized platform-neutral format (ex. XML). Lastly, these services often tend to be oriented towards use over a *network*.

As seen in Fig. 8, SOA consists of three entities: service provider, service requester, and service broker, and 3 operations: publish, find, and bind.

1. **Discovery Layer (publish)**: In the model, the service provider will first "publish" details of its service (description and location) to the service broker, who saves it to the Universal Description Discovery Integration (UDDI) registry. UDDI is an OASIS standard which provides a directory of services available from each service provider.
2. **Service Layer (find)**: The service requester queries the service broker with the "find" operation to find the service it is looking for, who will then return

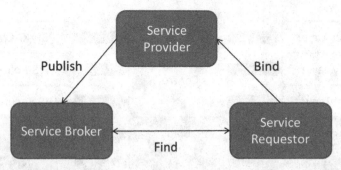

Fig. 8 Service Oriented Architecture (SOA)

the details of the service. This layer uses Web Services Description Language (WSDL), an XML-based W3C standard for describing network services as a set of endpoints operating on messages (W3C 2001).
3. **Message Layer (bind)**: Finally, the requester uses the connection details to "bind" to the provider and receive services. The message layer uses Simple Object Access Protocol (SOAP), an XML-based protocol for request and response messages in web services.

SOA is also been explored with mobile devices, with the mobile host acting as a service provider. The authors of (Fonseca et al. 2009) discuss the mobile host as a provider of services with SOA. It overviews the limitations with WS standards specifications on mobile cloud deployed services, as well as provide an architecture for supporting mobile clients in this environment. It has not been demonstrated in a real-life environment yet, although they are working on deploying it on Amazon EC2. Service-Oriented Architecture for Devices (SOA4D) (Fusion Forge n.d.) is an open-source initiative aimed at the development of service-oriented software components (SOAP, WS-*, etc.) to fit the needs of embedded devices. SOA4D implements Device Profile for Web Service (DPWS), a specification designed for secure Web service communications on resource-constrained devices, as further described below.

Device Profile for Web Services (DPWS)

When software is running on any device, the application will need to communicate with other services whether they are internal or external (over a network). The Device Profile for Web Services (DPWS) (OASIS 2009) follows the SOA framework for automatic device and service discovery for networked embedded devices. DPWS offers a standardized device representation of services on a network and this allows for access to a set of built-in services such as secure accessing of metadata and exchange services by utilizing WS protocols. In other words, DPWS defines a minimal set of implementation constraints to enable secure Web service messaging, discovery, description, control, and eventing on resource-constrained endpoints (OASIS 2009). The specification permits the definition of services for mobile devices considering the peer-to-peer direct communication between them that combine several devices as Service Oriented Architecture (SOA). DPWS allows sending secure messages to and from services, dynamically discovering a service, describing a service, subscribing to, and receiving events from a service.

In Web Services terms, a *profile* is a set of guidelines for how to use Web services technologies for a given purpose or application. Web services standards allow implementers to choose from a variety of message representations, text encodings, transport protocols, and other options, some of which are not interoperable. By constraining these decisions, profiles ensure that conforming implementations will work well together. DPWS is a profile developed by Microsoft and others for communication with and among networked devices and peripherals. The DPWS library for the.NET Micro Framework is not a full Web services implementation but a lightweight subset with only the functionality needed to support DPWS on a device (Microsoft 2007). DPWS was built on the foundation of existing web services (WS) and as such uses many common specifications such as XML, SOAP, WS-*, WSDL and Message Transmission Optimization Mechanism (MTOM). DPWS defines two main types of services that are run by devices: hosting services, and hosted services (Microsoft 2007). Devices can be DPWS clients (invoking hosted services on devices), servers (providing hosting services), or both. DPWS for the.NET Micro Framework supports devices in either role or both simultaneously. Hosted services are the services that the device has, and depends on their hosting service for discovery. Hosting services allow other devices to use, subscribe and obtain metadata of the given services. DPWS defines the extensions required for using services in mobile devices, taking in account their specific constraints. A DPWS enabled device has access to provided functionality such as: the discovery of other, utilizing WSDL to describe a Web service, service subscription, and secure sending of messages, given that the other device also utilizes DPWS.

The Web Services for Devices (WS4D) (Want 2006) framework is an extension of DPWS to bring SOA and Web services technology to industrial automation, home entertainment, automotive systems and telecommunication systems. There have been ongoing initiatives to connect internet technologies and web services to resource-constrained devices in ad-hoc networks while conserving interoperability. WS4D provides technologies for easy setup and management of network-connected devices in distributed embedded systems (Golatowski et al. n.d.). Araujo and Siqueira (Araujo und Siqueira 2009) used WS4D to implement a DPWS Device Service Bus (DSB), establishing a Device Tunnel to deal with virtual devices and services.

Pohlsen et al. (Pohlsen et al. 2009) present a plug-and-play architecture for connecting medical devices through DPWS, using WS-Discovery protocol. Unlike traditional Web service architectures, the authors propose using a WS-Discovery proxy server rather than a UDDI server, to better meet the requirements of resource constrained devices. Further, the work uses SOAP-over-UDP (User Datagram Protocol) for multicast messaging, as included in DPWS. El Kaed et al. (El Kaed et al. 2011) present an implementation to interoperably connect Universal Plug and Play

(UPnP) and DPWS smart home devices such as a TV, printer, and light bulb. DPWS does not support fine-grained security requirements, direct authentication between devices without a third party, and does not propose a comprehensive authorization concept (Unger et al. 2010). All of these works present the foundation technologies for this research work. To the best of our knowledge, there is no unified framework for enforcement for location privacy in mobile services for toy computing.

Section 3. Privacy and Access Control

Introduction to Privacy

When it comes to any information technology, privacy and security are at the core of ensuring that goals are achieved effectively and without compromise of personal data. The three concerns of security are confidentiality, integrity, and availability. Confidentiality means that access to information is restricted only to intended parties. Integrity means that data is accurate and consistent and has not been tampered with, while availability means that resources and data remain available when needed by the legitimate parties. A foundation of security is required for privacy.

Information privacy is defined by Hung and Cheng (Hung and Cheng, Privacy, 2009) as "an individual's right to determine how, when, and to what extent information about the self will be released to another person or to an organization." In particular, personally identifiable information is any type of information that can be linked to an individual, including their activities, preferences, history, conversations, etc. In a mobile environment, personally identifiable information is also likely to be gathered from context data, as described in the previous section. Information privacy goals can be achieved through privacy preserving mechanisms such as access control, privacy policies, and privacy preferences.

Walled Garden

In a toy computing environment, the concern is with the privacy of the user and that access to resources that can reveal context data are limited to the toy/game service application, and only used for purposes which comply with privacy regulations and are acceptable to the user. While the toy computing environment follows a BYOD model, it is required to identify a privacy preserving BYOD architecture. The Whitehouse has outlined three high-level means of implementing a BYOD program (Digital Services Advisory Group and Federal Chief Information Officers Council, United States of America 2012):

- **Virtualization**: Provide remote access to computing resources so that no data or corporate application processing is stored or conducted on the personal device;

- **Walled garden**: Contain data or corporate application processing within a secure application on the personal device so that it is segregated from personal data;
- **Limited separation**: Allow comingled corporate and personal data and/or application processing on the personal device with policies enacted to ensure minimum security controls are still satisfied.

In this context, a virtualized model would not be feasible or able achieve the privacy goals in a toy computing environment. However, privacy and security can be protected in a toy computing environment through the Walled Garden or Limited Separation approach. Walled Garden is a sandboxed and separated model which allows for processing to take place within a secure application which is separate from other applications and data. Limited separation allows the personal and corporate data and processing to comingle together, but enacts policies to protect the data and resources. Limited separation approach raises the issue of having a trusted mechanism for policy enforcement. To our best knowledge, not many research works are discussing the concept of Walled Garden.

Access Control

Access control is a security and privacy concept which aims to protect access to resources or data. The purpose of access control is to limit the actions or operations that a legitimate user can perform (Sandhu und Samarati 1994). There are two parts related to access control: the access decision, and the access enforcement. Access decisions can vary but the most basic are permit or deny. Access control decisions are made based on policies, for a variety of purposes. There are several different approaches to access control, including Mandatory Access Control (MAC) and Discretionary Access Control (DAC) (Open Web Application Security Project (OWASP), 2014), Role-based Access Control (RBAC) and Attribute-based Access Control (ABAC). In a DAC model, access decisions are based on the identity of users and/or membership in certain groups. Data owners are responsible for determining the type of access available to their resources. In MAC, sensitivity labels are assigned to users and resources. In this model, users are granted or denied access based on their security clearance and the label associated with the resource. Further, RBAC determines access to resources/data based on the role of the subject. Attribute-based access control makes access decisions based on attributes associated with subjects and objects. Access control

There are different types of policies which an access decision can be based on, e.g. privacy policies and security policies. Security policies are focused on maintaining confidentiality, integrity and availability of resources, while privacy policies are concerned with how and why data is used/shared/stored/etc. Privacy policies are the focus of access control decisions for the purpose of this work, and will be further described in the next section.

Privacy Policies and Preferences

Privacy policies describe an enterprise's data practices. This includes a description of what information is collected from users, what the information will be used for, how long it will be held, if/how the information will be shared to third parties, how long the information will be retained, etc. Consent is given by the user either implicitly or explicitly. Often, consent is implied just by using the services. Explicit consent can be given if the user is required to click "I agree" in regards to the privacy policy terms and conditions in order to receive services. Privacy policies are used for a company to outline their privacy practices relating to collection, use, retention, and sharing practices. Privacy preferences allow the user to create a set of rules to express how they wish their information to be managed.

Human Readable Policies

Privacy policies are often provided to their users in natural language. Mobile applications often provide privacy policies to their users in this format. The purpose of these privacy policies is to provide the user with the details on why and how their information is collected while they are using the mobile application. As an illustration, the following is the *Furby Boom!* App Privacy Policy, available online or through the app:

> Hasbro may collect non-personally identifiable information from devices that have installed a Hasbro app. This information is used to deliver services requested by users, such as content and updates within the app, as well as to support the internal operations of the app. For more information about the app, please contact us at http://hasbro-new.custhelp.com/ (Hasbro 2013)

This policy is available before installing the application, and provides the user with an idea of what type of information is collected, and what the purpose is for its collection. This human-readable privacy policy is short and in simple terms, however it does not provide any detail on what information is actually collected, or how exactly it is used.

There are several concerns with how privacy policies are used in practice. In the case where a privacy policy is provided, the majority of users find them too complicated or long to read. Alternatively, as in the case of the *Furby Boom!* Privacy Policy, they can also be too vague. Human-readable privacy policies have a lot of limitations, some of which can be improved through the use of machine-readable policies.

XML and Machine Readable Policies

Structured policy languages allow for automated enforcement of privacy policies and access decisions. A privacy policy language supports access constraints (e.g. which subject can perform which action on which resource), as well as a description of access conditions. Policy languages must be platform independent, and able to integrate with the language used for access control policies (Anderson 2006). Privacy policies can be expressed in eXtensible Markup Language (XML) (W3C 2015) through policy assertion languages. XML is a flexible markup language used to describe data. XML is both human readable and machine readable, and many APIs have been developed for processing XML data. Various languages and tools have been developed for the specification of privacy policies and preferences based on XML, including P3P, EPAL, XACML, and WS-policy.

Platform for Privacy Preferences (P3P)

Machine-readable privacy policy frameworks differ from human-readable ones. With machine-readable privacy policies, it allows the user to have more control over what information is collected and stored. Platform for Privacy Preferences (P3P) is a privacy policy framework created by the World Wide Web Consortium (W3C), based on XML designed to help end users manage their privacy while navigating websites that have differing privacy policies. User's privacy preferences are expressed using APPEL, A P3P Preference Exchange Language P3P also enables Websites to express their privacy practices in a standard format that can be retrieved automatically and interpreted easily by users of P3P browsers (Wenning 2007). P3P addresses user concerns about the type and number of data gathered by websites. At its most basic, any website that collects user information must clearly declare the reasons for the data collection, how it plans to use the information, and the amount of time it will retain the information. When using a P3P-compliant browser, cookies will be accepted, bypassed or denied depending on the previously mentioned user preferences. The user receives an alert when any privacy concerns arise and can override the previously set privacy level if they wish.

While P3P was primarily designed for Web sites, it has been the focus of many future directions including Web services and mobile services. In (Olurin et al. 2012), adaptation for the mobile environment is noted as a prominent future direction for P3P. Some major research questions are also addressed in this paper, including: how to create mobile-based privacy user agents that can communicate compact privacy policies of mobile web sites or applications to users, and how to delegate automatic access control privileges to mobile applications and websites based on user defined privacy preferences. Some concerns with moving P3P to the mobile environment, as outlined in (WAP-W3C 2000), include performance, security of the policies, extending P3P vocabulary for the mobile environment, and adapting the user interface for use on small mobile devices.

The traditional approach to P3P has several shortfalls in terms of enforcement. (Cranor, P3P is Dead, Long Live P3P! 2012b) reiterates how P3P has not been strongly embraced in practice. Popular websites such as Google and Facebook have published P3P "compact policies" (Cranor, Internet Explorer Privacy Protections also Being Circumvented by Facebook, and Many more 2012a). These policies state in human-readable code "this is not a P3P policy," while in practice, the system interprets it as a valid policy. In these situations, websites are able to technically comply with requirements but do not provide any actual privacy enforcement.

The authors of (Reay et al. 2009) performed an analysis on over 3000 P3P policies from 100,000 web sites to determine the relationship of privacy policies compared to legal requirements. The results of this study indicated that the surveyed website privacy policy statements had a widespread lack of adherence to legal mandates. Another report from the Canadian Internet Policy and Public Interest Clinic (Seligy und Lawson 2006) found similar results in a survey of 72 Canadian websites showing widespread noncompliance with PIPEDA. Many businesses are not taking necessary steps to preserve the privacy of their users. Another issue is faced by international web service companies (e.g. Google and Yahoo), who have difficulty with privacy regulation while they are required to address a multitude of different or conflicting international privacy laws and jurisdictions that must be negotiated (Reay et al. 2009).

Enterprise Privacy Authorization Language (EPAL)
EPAL (Ashley et al. Enterprise Privacy Authorization Language (EPAL 1.2) 2003a) is another XML-based privacy policy language by W3C member IBM, designed to formalize internal privacy practices of an enterprise. EPAL is more suitable than P3P to express internal privacy policies that can be enforced by the enterprise's privacy management system. EPAL allows an enterprise to define its own list of data categories, data users, purposes, and actions, whereas P3P is limited to a predefined list (Ashley et al. The Enterprise Privacy Authorizsation Language (EPAL)–How to Enforce Privacy Throughout an Enterprise 2003b).

eXtensible Access Control Markup Language (XACML)
eXtensible Access Control Markup Language (XACML) (OASIS 2013) is an OASIS standard for access control language and architecture. The policy language uses the XML standard to define the policy and access control decision request and response. When there is an access request, an authorization decision/response can then be made based on the policy. XACML supports both centralized and decentralized policy management. XACML architecture uses the IETF Abstract Model for Policy Enforcement, which is further described in Sect. 2.3.5.

XACML specifies an abstract format for the authorization decision request as a description of the attempted resource access in terms of attributes (Anderson 2006). An XACML attribute is associated with one of four classes: Subject, Resource, Action, and Environment. Subject is the entity who is sending the access request, Resource is the resource that is to be accessed, and Action is the action to be performed on the resource (e.g. read or write). The environment attribute describes an additional characteristic of the request such as time of day.

The use of XACML has been widely adopted in Web services (Anderson 2006). A comparison between EPAL and XACML by Anderson (Anderson 2006), has recommended XACML for its functionality and flexibility. Lastly, Geospatial eXtensible Access Control Markup Language (GeoXACML) (Open Geospatial Consortium 2005) is an extension to XACML Version 2.0. by the Open Geospatial Consortium designed to control access to geospatial information. GeoXACML supports four types of functions: topological, geometric, bag & set, and conversion to manage geospatial information.

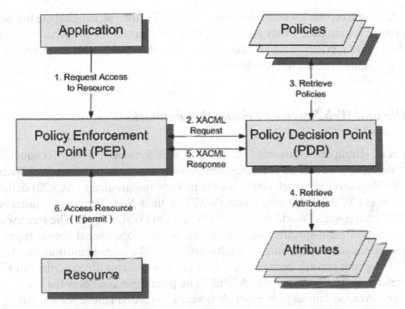

Fig. 9 IETF abstract model for policy enforcement. (Adapted from Waters et al. 1999)

Abstract Model for Policy Enforcement

A privacy policy alone does not guarantee that the policies will actually be enforced. This brings us onto the Abstract Model for Policy Enforcement proposed by IETF (terminology (Westerinen et al. 2001), model (Yavatkar et al. 2000)) and ISO (Open Systems Interconnection, 1966). This model has been used for policy enforcement for privacy policy languages such as EPAL and XACML.

Referring to Fig. 9, access control decisions are made by the Policy Decision Point (PDP), and enforced by the Policy Enforcement Point (PEP). When an application requests access to a resource, it sends the request to the PEP, which forwards the request to the PDP. The PDP then retrieves the policies and attributes to determine if the request complies. The PDP will make a decision and send a Permit or Deny response back to the PEP. The PEP will enforce the decision accordingly, providing access to the resource if permitted.

Section 4. Related Works in Privacy, Mobile and Location Services

There exist a number of previous works in the fields of privacy, mobile services, and location-based services. To the best of our knowledge, no previous work exists which identifies a framework for privacy exclusively in the context of toy computing, and especially with a focus on location. Further, although there has been work on the topic, there exists no widely accepted framework for privacy for any type of mobile services. This section will provide the reader with an overview of existing literature related to these topics.

Mobile and Web Services Privacy Frameworks

Hung et al. (Hung et al. Towards Standardized Web Services Privacy Technologies 2004) describe a vocabulary-independent privacy authorization language framework for Web services which addresses the privacy requirements (AC020) defined by the World Wide Web Consortium (W3C) in their Web Services Architecture (WSA) Requirements (World Wide Web Consirtium (W3C) 2004). The framework recommends domain-specific vocabularies to be developed for different types of business applications (e.g. finance, healthcare, etc.). The authors introduce a protocol for enforcing privacy policies, in which privacy policies are described in P3P, and preferences exchange rules in APPEL. The paper also considers the use of privacy authorization language in other Web services-related languages such as WS-Policy, WS-Security, and WS-Privacy.

Access control is another area in which privacy is becoming more important. Traditional access control mechanisms such as discretionary access control (DAC), mandatory access control (MAC), and role based access control (RBAC) are not generally designed to accommodate privacy (Ferraiolo und Kuhn 1992), however some recent RBAC extensions have been introduced with a privacy-focused objective (Ni et al. 2007). Context- and location-based access control models have also been proposed (Seifert et al. 2009), where certain services and data can only be accessed in a certain context/location. This is especially useful in a BYOD scenario where users wish to separate work from personal activities depending on their context.

A lattice-based privacy aware access control (LPAAC) model is described in (Ghazinour und Barker 2013), in which data provider and collector privacy preferences are accommodated and enforced. This model allows the data collector to identify their privacy policies for purpose, visibility, granularity, and retention of data in terms of minimal acceptance limit (MinAL) and maximal acceptance limit (MaxAL). The data provider can then review the privacy policies and select their own preferences within the range, allowing them to receive services from the data collector while still being in control of their data. This paper also identifies the importance of enforcement, and provides an algorithm based on the above for determining the access decision to be enforced by the system it is being implemented on. The authors have also implemented their model using P3P (Ghazinour und Barker 2011).

ipShield, introduced by (Chakraborty et al. 2013b), is a privacy-aware framework designed to quantify an adversary's knowledge regarding the user's context and obscure it before sharing. This framework does not depend on the user being anonymous, but instead focuses on choosing which data to share. It identifies several information disclosure systems, each corresponding to a specific privacy-utility tradeoff. Also introduces privacy mechanisms designed to realize those tradeoff points. Chakraborty et al. (Chakraborty et al. 2013a) propose a framework for protecting data against unwanted inferences. This technique involves a white list of inferences that are desirable and provide utility, as well as a black list for unwanted inferences that should be kept private. From there, the authors attempt to define how much the recipient can infer from shared data based on utility-privacy parameters. They identify bounds on the parameters and provide mechanisms for achieving the bounds.

Location Privacy Techniques

Various techniques have been used in attempt to preserve the privacy of a user's location. Different approaches could involve or not involve a trusted third party (Solanas et al. 2008). Some approaches include degrading the quality of location information (obfuscation) (Duckham und Kulik 2005) (Ardagna et al. 2007), creating fake location points (Taha und Shen 2013), uncertainty (Cheng und Prabhakar 2004) (Merrill et al. 2013), pseudonyms (Jorns et al. 2005), encryption (Fang et al. 2011) (Ashouri-Talouki und Baraani-Dastjerdi 2012), and k-anonymity (Gedik und Liu 2005). Policy-based access control is another technique which is used to decide whether a requesting subject can perform a given action on a data object. Various approaches for context-aware access control have been explored, which can also be used to preserve location privacy (Riboni et al. 2008).

IETF RFC6280 by Barnes et al. (Barnes et al. 2011) presents Geopriv, an architecture for location and location privacy in Internet applications. Geopriv is an Internet Best Current Practice, which enables users to express preferences for the disclosure of their location information. For example, the user can make a rule that their location is not to be disclosed beyond the intended recipient. This architecture binds the privacy rules to the data so that receiving entities are informed of when their data is shared to other parties.

Section 5. Chapter Summary

In this chapter, we provided a background on the concept of toy computing, including the concepts of mobile services and physical computing. Next, we established a foundation on privacy in this context, including a description of XML-based privacy policy assertion languages including P3P and XACML. Finally, we provided an overview of some related works on mobile/web services privacy frameworks and location privacy.

References

Anderson, A. H. (2006). A Comparison of Two Privacy Policy Languages: EPAL and XACML. *Proceedings of the 3rd ACM Workshop on Secure Web Services*, (pp. 53–60). New York, NY.

Andreae, H., Andreae, P., Low, J., & Brown, D. (2014). A Study of Auti: A Socially Assistive Robotic Toy. *IDC '14 Proceedings of the 2014 Conference on Interaction Design and Children* (pp. 245–248). New York, NY, USA: ACM.

Android. (2015). *Location Strategies*. (Android Developer) Retrieved February 2015, from http://developer.android.com/guide/topics/location/strategies.html

Araujo, G. M., & Siqueira, F. (2009). The Device Service Bus: A Solution for Embedded Device Integration through Web Services. *Proceedings of the 2009 ACM Symposium on Applied Computing* (pp. 185–189). New York, NY: ACM.

Ardagna, C. A., Cremonini, M., Damiani, E., De Capitani di Vimercati, S., & Samarati, P. (2007). Location Privacy Protection Through Obfuscation-based Techniques. In *Lecture Notes in Computer Science: Data and Applications Security* (Vol. 4602, pp. 47–60). Redondo Beach, California, USA: Springer Berlin Heidelberg.

Ashley, P., Hada, S., Karjoth, G., Powers, C., & Schunter, M. (2003a, November 10). *Enterprise Privacy Authorization Language (EPAL 1.2)*. Retrieved March 2015, from http://www.w3.org/Submission/2003/SUBM-EPAL-20031110/

Ashley, P., Hada, S., Karjoth, G., Powers, C., & Schunter, M. (2003b). *The Enterprise Privacy Authorizsation Language (EPAL)—How to Enforce Privacy Throughout an Enterprise*. Retrieved March 2015, from http://www.w3.org/2003/p3p-ws/pp/ibm3.html

Ashouri-Talouki, M., & Baraani-Dastjerdi, A. (2012). Homomorphic Encryption to Preserve Location Privacy. *International Journal of Security and Its Applications, 6*(4), 183–190.

Baber, C., & Westmancott, O. (2004). Social networks and mobile games: the use of bluetooth for a multiplayer card game. *6th International Conference on Human Computer Interaction with Mobile Devices and Services*, (pp. 98–107). Glasgow, Scotland.

Barnes, R., Lepinski, M., Cooper, A., Morris, J., Tschofenig, H., & Schulzrinne, H. (2011). *An Architecture for Location and Location Privacy in Internet Applications*. IETF. Retrieved from http://tools.ietf.org/html/rfc6280

Beckett, P. (2014). BYOD—Popular and Problematic. *Network Security, 2014*(9), 7–9.

Booyah. (2015). *iTunes—MyTown2*. Retrieved February 2015, from https://itunes.apple.com/app/mytown-2/id442345455

Chakraborty, S., Bitouze, N., Srivastava, M., & Dolocek, L. (2013a). Protecting Data Against Unwanted Inferences. *Proceedings of the 2013 IEEE Information Theory Workshop*. Seville, Spain.

Chakraborty, S., Raghavan, K., Johnson, M., & Srivastava, M. (2013b). *A Framework for Context-Aware Privacy of Sensor Data on Mobile Systems*. The Fourteenth Workshop on Mobile Computing Systems and Applications (ACM HotMobile2013) (pp. 1–6, Article 11). New York, USA: ACM.

Cheng, R., & Prabhakar, S. (2004). Using Uncertainty to Provide Privacy-Preserving and High-Quality Location-Based Services. *Workshop on Location Systems Privacy and Control (mobileHCI'04)*. Glasgow, Scotland.

Cherubini, M., de Oliveira, R., Hiltunen, A., & Oliver, N. (2011). Barriers and bridges in the adoption of today's mobile phone contextual services. *MobileHCI '11* (pp. 167–176). Stockholm, Sweden: ACM.

ChineseCUBES. (2014). *AR Cubes*. Retrieved from https://www.chinesecubes.com/ar_cubes

Cranor, L. (2012a, February 18). *Internet Explorer Privacy Protections also Being Circumvented by Facebook, and Many more*. Retrieved October 2013, from TechPolicy.com: http://www.techpolicy.com/Cranor_InternetExplorerPrivacyProtectionsBeingCircumvented-by-Google.aspx

Cranor, L. (2012b, December 3). *P3P is Dead, Long Live P3P!* Retrieved October 2013, from http://lorrie.cranor.org/blog/2012/12/03/p3p-is-dead-long-live-p3p/

Cumulonimbus. (2015). *PressureNet*. (Google Play) Retrieved February 2015, from https://play. google.com/store/apps/details?id=ca.cumulonimbus.barometernetwork&feature=nav_result#-?t=W251bGwsMSwxLDMsImNhLmN1bXVsb25pbWJ1cy5iYXJvbWV0ZXJuZXR3b3b3JrIl0.

Dewri, R., Annadata, P., Eltarjaman, W., & Thurimella, R. (2013). Inferring Trip Destinations from Driving Habits Data. *Workshop on Privacy in the Electronic Society*. Berlin, Germany.

Dey, A. K., & Abowd, G. D. (1999). *Towards a Better Understanding of Context and Context-Awareness*. Georgia Institute of Technology, College of Computing.

D'Hooge, H., & Goldstein, M. (2001). History of the Smart Toy Lab and Intel Play Toys. *Intel Technology Journal, 2001*(Q4).

Digital Services Advisory Group and Federal Chief Information Officers Council, United States of America. (2012, August 23). *Bring Your Own Device*. Retrieved September 2014, from http://www.whitehouse.gov/digitalgov/bring-your-own-device

Disterer, G., & Kleiner, C. (2013). BYOD Bring Your Own Device. *Procedia Technology, 2013*(9), 43–53.

Duckham, M., & Kulik, L. (2005). A Formal Model of Obfuscation and Negotiation for Location Privacy. In H. G. al. (Ed.), *Pervasive Computing* (Vol. 3468, pp. 152–170). Munich, Germany: Springer-Verlag Berlin Heidelberg.

Duri, S., Cole, A., Munson, J., & Christensen, J. (2001). An approach to providing a seamless end-user experience for location-aware applications. *1st International Workshop on Mobile Commerce, 86*(4), 20.

El Kaed, C., Denneulin, Y., & Ottogalli, F.-G. (2011). Dynamic Service Adaptation for Plug and Play Device Interoperability. *Proceedings of the 7th International Conference on Network and Services Management*.

Facebook. (2015). *Facebook Mobile*. (Facebook) Retrieved February 2015, from https://www.facebook.com/mobile/

Fang, S.-H., Lai, W.-J., & Liang, Y.-C. (2011). An Encryption-Based Approach for Protecting Privacy in Network-Based Location Systems. *IEEE 2011 International Conference on Machine Learning and Cybernetics (ICMLC)* (pp. 377–380). Guilin: IEEE.

Ferraiolo, D., & Kuhn, R. (1992). Role-based Access Control. *Proceedings of the 15th National Computer Security Conference*, (pp. 1–11).

Fonseca, J., Abdelouahab, Z., Lopes, D., & Labidi, S. (2009). A Security Framework for SOA Applications on Mobile Environment. *International Journal of Network Security & ITS Applications, 1*(3), 90–107.

Fusion Forge. (n.d.). *Welcome to the SOA4D Forge*. Retrieved September 2014, from https://forge.soa4d.org/

Futuresight. (2011). *User Perspectives on Mobile Privacy—Summary of Research Findings*. GSMA.

Gartner. (2014). *Key Challenges in BYOD*. Retrieved September 2014, from http://www.gartner.com/technology/topics/byod.jsp

Gedik, B., & Liu, L. (2005). Location Privacy in Mobile Systems: A Personalized Anonymization Model. *25th IEEE International Conference on Distributed Computing Systems* (pp. 620–629). Columbus, OH: IEEE.

Ghazinour, K., & Barker, K. (2011). Capturing P3P Semantics Using and Enforceable Lattice-based Structure. *Proceedings of the 4th International Workshop on Privacy and Anonymity in the Information Society*, (p. 4). New York, NY, USA.

Ghazinour, K., & Barker, K. (2013). A Privacy Preserving Model Bridging Data Provider and Collector Preferences. *Proceedings of the Joint EDBT/ICDT 2013 Workshops* (pp. 174–178). New York, NY, USA: ACM.

Golatowski, F., Bobek, A., & Zeeb, E. (n.d.). *Web Services for Devices—About*. (WS4D) Retrieved September 2014, from http://ws4d.e-technik.uni-rostock.de/about/

Google. (n.d.). *Sensors Overview*. (developer.android.com) Retrieved September 2014, from http://developer.android.com/guide/topics/sensors/sensors_overview.html

Gross, M. D., & Eisenberg, M. (2007). Why Toys Shouldn't Work "Like Magic": Children's Technology and the Values of Construction and Control. *The First IEEE International Workshop on*

Digital Game and Intelligent Toy Enhanced Learning (DIGITEL '07). Jhongli, Taiwan: IEEE Computer Society.

Gu, T., Pung, H., & Zhang, D. (2004). A Middleware for Building Context-Aware Mobile Services. *IEEE Vehicular Techology Conference. 5*, pp. 2656–2660. IEEE.

Hasbro. (2013). *Furby Boom*. Hasbro. Retrieved November 2013, from http://www.hasbro.com/furby/en_CA/

Hinske, S., & Langheinrich, M. (2007). Managing Augmented Toy Environments—A New Perspective for Smart Space Management. *Proceedings of the 4th International Workshop on Managing Ubiquitous Communications and Services (MUCS)*. Munich, Germany.

Hinske, S., Langheinrich, M., & Lampe, M. (2008). Towards Guidelines for Designing Augmented Toy Environments. *Designing Interactive Systems (DIS) 2008*. Cape Town, South Africa: ACM.

Hung, P. C., & Cheng, V. S. (2009). Privacy. In *Encyclopedia of Database Systems* (pp. 2136–2137). Springer.

Hung, P. C., Ferrari, E., & Carminati, B. (2004). Towards Standardized Web Services Privacy Technologies. *Proceedings of the IEEE International Conference on Web Services (ICWS'04)*. San Diego, CA.

Instagram. (2015). *Instagram*. (Instagram) Retrieved February 2015, from http://instagram.com/

Jorns, O., Jung, O., Gross, J., & Bessler, S. (2005). A Privacy Enhancement Mechanism for Location Based Service Architectures Using Transaction Pseudonyms. In *Lecture Notes in Computer Science: Trust, Privacy, and Security in Digital Business* (Vol. 3592, pp. 100–109). Copenhagen, Denmark: Sprinter-Verlag Berlin Heidelberg.

Kaasinen, E. (2003, May). User Needs for Location-Aware Mobile Services. *Personal and Ubiquitous Computing, 7*(1), 70–79.

Lee, S., & Doh, Y. Y. (2013). iSpy: RFID-Driven Language Learning Toy Integrating Living Environment. *CHI '13 Proceedings from the 2013 International Conference on Interaction Design and Children* (pp. 697–702). Paris, France: ACM.

Luckin, R., Connolly, D., Plowman, L., & Airey, S. (2003). Children's Interactions with Interactive Toy Technology. *Journal of Computer Assisted Learning, 19*, 165–176.

MEF. (2013). *MEF Global Privacy Report 2013*. MEF.

Merriam-Webster. (n.d.). *Location*. Retrieved January 2015, from http://www.merriam-webster.com/dictionary/location

Merrill, S., Basalp, N., Biskup, J., Buchmann, E., Clifton, C., Kuijpers, B., … Savas, E. (2013). Privacy Through Uncertainty in Location-Based Services. *2013 IEEE 14th International Conference on Mobile Data Management* (pp. 67–72). Milan, Italy: IEEE.

Microsoft. (2007). *Intorducing Devices Profile for Web Services*.

National Geographic. (n.d.). *Encyclopedic Entry: Location*. Retrieved January 2015, from http://education.nationalgeographic.com/education/encyclopedia/location/?ar_a=1

Ni, Q., Trombetta, A., Bertino, E., & Lobo, J. (2007). Privacy-Aware Role Based Acces Control. *SACMAT '07: Proceedings of the 12th ACM Symposium on Access Control Models and Technologies* (pp. 41–50). France: ACM.

OASIS. (2009, July). *OASIS Devices Profile for Web Services*. Retrieved December 2013, from http://docs.oasis-open.org/ws-dd/ns/dpws/2009/01

OASIS. (2013). *eXtensible Access Control Markup Language (XACML) Version 3.0*. OASIS. Retrieved January 2015, from http://docs.oasis-open.org/xacml/3.0/xacml-3.0-core-spec-os-en.pdf

Olurin, M., Adams, C., & Logrippo, L. (2012). Platform for Privacy Preferences (P3P): Current Status and Future Directions. *2012 Tenth Annual International Conference on Privacy, Security and Trust (PST)*, (pp. 217–220). Paris, France.

Open Geospatial Consortium. (2005, December 15). *GeoXACML*. Retrieved December 2014, from https://geoxacml.secure-dimensions.com/

Open Systems Interconnection. (1966). Information Technology—Security Frameworks for Open Systems: Access Control Framework.

Open Web Application Security Project (OWASP). (2014). *Access Control Cheat Sheet*. OWASP. Retrieved from https://www.owasp.org/index.php/Access_Control_Cheat_Sheet

Pandit, A. A., & Kumar, A. (2012). Conceptual Framework and a Critical Review for Privacy Preservation in Context Aware Systems. *IEEE 2012 International Conference on Cyber-Enabled Distributed Computing and Knowledge Discovery* (pp. 435–442). Sanya, China: IEEE.

Patil, S., Norcie, G., Kapadia, A., & Lee, A. J. (2012). Reasons, Rewards, Regrets: Privacy Considerations in Location Sharing as an Interactive Practice. *Symposium on Usable Privacy and Security*. Washington, D.C.

Plowman, L., & Luckin, R. (February 2004). Interactivity, Interfaces, and Smart Toys. *Computer*, 98–100.

Pohlsen, S., Schlichting, S., Strahle, M., Franz, F., & Werner, C. (2009). A Concept for a Medical Device Plug-and-Play Architecture based on Web Services. *ACM SIGBED Review—Special Issue on the 2nd Joint Workshop on High Confidence Medical Devices, Software, and Systems (HCMDSS) and Medical Device Plug-and-Play (MD PnP) Interoperability, 6*(2), Article 6.

Pura, M. (2005). Linking perceived value and loyalty in location-based mobile services. *Managing Services Quality, 15*(6), 509–538.

Reay, I., Dick, S., & Miller, J. (2009). A Large-Scale Empirical Study of P3P Privacy Policies: Stated Actions vs. Legal Obligations. *ACM Transactions on The Web, 3*(2), 6:1–6:34.

Riboni, D., Pareschi, L., & Bettini, C. (2008). Privacy in Georeferenced Context-aware Services: A Survey. *Privacy in Location-Based Applications (PiLBA'08)*, (pp. 24–43). Malaga, Spain.

Rovio. (2015). *Angry Birds*. Retrieved February 2015, from http://www.rovio.com/en/our-work/games/view/1/angry birds

Saha, D. (2003). Pervasive Computing: A Paradigm for the 21st Century. *Computer, 36*(3), 25–31.

Sandhu, R., & Samarati, P. (1994). Access Control: Principles and Practice. *IEEE Communications Magazine, 1994*(September), 40–48.

Schilit, B., Adams, N., & Want, R. (1994). Context-Aware Computing Applications. *WMCA '94* (pp. 85–90). Washington, DC, USA: IEEE Computer Society.

Schmidt, A. (2005). Interactive Context-Aware Systems Interacting with Ambient Intelligence. In G. Riva, F. Vatalaro, F. Davide, & M. Alcaniz (Eds.), *Ambient Intelligence* (pp. 159–178). IOS Press.

Schmidt, A., Beigle, M., & Gellersen, H. W. (1999). There is more to context than location. *Computer & Graphics Journal, 23*(6), 893–902.

Seifert, J., De Luca, A., & Conradi, B. (2009). A Context-Sensitive Security Model for Privacy Protection on Mobile Phones. *Proceedings of the 11th International Conference on Human-Computer Interaction with Mobile Devices and Services*. New York, NY.

Seligy, J., & Lawson, P. (2006). *Compliance with Canadian Data Protection Laws: Are Retailers Measuring Up?* Ottawa: Canadian Internet Policy and Public Interest Clinic.

Sheth, A., Anantharam, P., & Henson, C. (2013). Physical-Cyber-Social Computing: An Early 21st Century Approach. *Intelligent Systems, IEEE, 28*(1), 78–82.

Six to Start. (n.d.). *Zombies, Run! 3*. (Six to Start) Retrieved September 2014, from https://www.zombiesrungame.com/

Solanas, A., Domingo-Ferrer, J., & Martinez-Balleste, A. (2008). Location Privacy in Location-Based Services: Beyond TTP-based Schemes. *Privacy in Location-Based Applications (PiLBA'08)*, (pp. 12–23). Malaga, Spain.

Sphero. (2014). *Sphero*. Retrieved August 2014, from http://www.gosphero.com

Taha, S., & Shen, X. (2013). A Physical-Layer Location Privacy-Preserving Scheme for Mobile Public Hotspots in NEMO-Based VANETs. *IEEE Transactions on Intelligent Transportation Systems, 14*(4), 1665–1680.

Tech4Kids. (2013). *Tek Recon*. (Tech4Kids) Retrieved August 2014, from http://www.tekrecon.com/

TIME and Qualcomm. (2012, July). *Your Wireless Life: Results of TIME's Mobility Poll*. Retrieved November 2013, from http://content.time.com/time/interactive/0,31813,2122187,00.html

Toymail Co. LLC. (2014). *Toy Mail*. (Toymail Co. LLC) Retrieved from http://www.toymail.co/

Unger, S., Zeeb, E., Golatowski, F., Grandy, H., & Timmermann, D. (2010). Extending the Devices Profile for Web Services for Secure Mobile Device Communication. *4th International Workshop on Trustworthy Internet of People, Things & Services at the Internet of Things Conference*. Tokyo, Japan. Retrieved from http://webcache.googleusercontent.com/search?q=cache:ArR-T87Qq_8 J:www.imd.uni-rostock.de/veroeff/DPWS-DA-paper.pdf

W3C. (2001). *Web Services Description Language (WSDL) 1.1*. W3C. Retrieved from www.w3.org/TR/wsdl

W3C. (2004). *Web Services Architecture*. Retrieved 2014, from http://www.w3.org/TR/ws-arch

W3C. (2015). Extensible Markup Language (XML). W3C. Retrieved from http://www.w3.org/XML

Want, R. (2006). An Introduction to RFID Technology. *IEEE Pervasive Computing, 5*(1).

WAP-W3C. (2000). *Report from WAP-W3C Joint Workshop on Mobile Web Privacy*. Munich, Germany: W3C.

Waters, G., Wheeler, J., Westerinen, A., Rafalow, L., & Moore, R. (1999). Policy Framework Architecture. IETF. Retrieved from http://tools.ietf.org/html/draft-ietf-policy-arch-00

Wenning, R. (2007, November 20). *Platform for Privacy Preferences (P3P) Project: Enabling Smarter Privacy Tools for the Web*. Retrieved December 2013, from W3C: http://www.w3.org/P3P/

Westerinen, A., Schnizlein, J., Strassner, J., Scherling, M., Quinn, B., Herzog, S., ... Waldbusser, S. (2001, November). Terminology for Policy-Based Management. IETF RFC 3198. Retrieved from http://www.ietf.org/rfc/rfc3198.txt

Westeyn, T. L., Abowd, G. D., Starner, T. E., Johnson, J. M., Presti, P. W., & Weaver, K. A. (2012). Monitoring Children's Developmental Progress using Augmented Toys and Activity Recognition. *Personal Ubiquitous Computing, 2012*(16), 169–191.

Whalen, T. (2011). Mobile Devices and Location Privacy: Where do we go from Here? *IEEE Security & Privacy, 9*(6), 61–62.

Woollaston, V. (2014, February 20). *Step Aside Ted, There's A New Talking Teddy in Town: WikiBear Connects to the Web to Chat, Answer Questions and Tell Jokes*. Retrieved from DailyMail: http://www.dailymail.co.uk/sciencetech/article-2564015/Step-aside-Ted-theres-new-talking-teddy-town-WikiBear-connects-web-chat-answer-questions-tells-jokes.html

World Economic Forum. (2011). *Personal Data: The Emergence of a New Asset Class*. World Economic Forum.

World Wide Web Consirtium (W3C). (2004, February 11). *Web Services Architecture Requirements*. Retrieved from http://www.w3.org/TR/wsa-reqs/

Yavatkar, R., Pendarakis, D., & Guerin, R. (2000, January). A Framework for Policy-Based Admission Control. IETF RFC2753. Retrieved from http://www.ietf.org/rfc/rfc2753.txt

Yelp. (2015). *Yelp Mobile*. (Yelp) Retrieved February 2015, from http://www.yelp.ca/yelpmobile

zomato. (2015). *Urbanspoon*. (zomato) Retrieved February 2015, from http://www.urbanspoon.com/

Mobile Computing Toys: Marketing Challenges and Implications

Tirtha Dhar and Terry Wu

Abstract Global sales of traditional toys such as dolls, action figures, and role-play games are either flat or shrinking. One of the main reasons for the slow or negative growth of sales is the increase in internet and console-based video gaming across multiple devices and platforms. Faced with this disruptive force, traditional toy makers are increasingly trying to develop toys that incorporate interactivity through software integration and mobile interconnectivity. Though the idea of creating this synergy between traditional toy hardware and emerging mobile/internet connectivity is a no-brainer, the success of such integration depends on understanding the market fundamentals of hardware and software complementarity. The responses of the toy industry thus far have been ineffective in stopping the decline of sales. This paper takes a closer look at the issue by conducting a brief survey of the traditional toy industry and trying to identify the key drivers of success and failure of such toys. We then use a few tools of marketing research to explain how traditional toy makers can leverage information and mobile technologies to re-establish themselves in a sustainable growth path.

Keywords Mobile Computing Toys · Marketing · Toy Market

Introduction

Children's toys have become increasingly sophisticated over the years, with a growing shift from simple physical products to toys that engage the digital world. Consumers attracted to products in the emerging digital medium are demanding more convergence and interconnectedness. Toy makers are seizing this opportunity to develop products that combine the characteristics of traditional toys (such as dolls, cars, building blocks, etc.) with computing software and hardware. Many of these toy makers are also leveraging emerging trends in mobile computing and consumer

T. Wu (✉) · T. Dhar
Faculty of Business and IT, University of Ontario Institute of Technology, Oshawa, Canada
e mail: terry.wu@uoit.ca

© Springer International Publishing Switzerland 2015
P. C. K. Hung (ed.), *Mobile Services for Toy Computing*, International Series on
Computer Entertainment and Media Technology, DOI 10.1007/978-3-319-21323-1_3

ownership of mobile devices in order to enhance the experience of users and to increase toy sales.

In this study, we focus on this emerging mobile computing toy category that combines traditional toys with mobile computing functionalities. For the purpose of our discussion, we will term it "the mobile computing toy" category (hereafter MCT). For traditional toy companies, as the global growth of traditional toys slows, this new product category holds immense promise to restart this growth engine. Similarly, for mobile device makers, this new product category provides an opportunity to apply their know-how to a completely different, unexplored market.

The purpose of this study is to discuss marketing issues and implications of MCTs. The objective is to outline some of the key challenges facing MCTs and offer suggestions for overcoming some of these challenges. The paper is organized as follows. First we define and outline a history of toys. Next we provide our analysis of the current state of toy marketing. Based on our understanding of the role of toys, we then delineate their key characteristics. Finally, we discuss core challenges facing businesses that may develop new toys in the MCT category and then provide relevant insights into potential marketing strategies.

In terms of providing strategic insights, we should mention that from the outset, we are surprised by the lack of extant literature addressing MCT marketing and toys in general. So, rather than relying on existing literature to structure our arguments, we will rely on analytical marketing and business tools to provide insights into building successful MCT businesses. With this goal in mind, we start by outlining some of the key elements found in the most successful toys. We then use analytical tools from marketing and examples from other successful business integrations to develop our strategic recommendations.

A Brief History of Toys

A toy is defined as "a small representation of something familiar, as an animal or person, for children or others to play with or plaything."[1] The purpose of toys is to entertain, develop motor skills, improve the process of imagination, and enhance social skill formation. For these reasons, decisions to purchase toys depend not only on children, but also on their parents. So, for any toy to be successful, it has to have the ability to influence parents and children at the same time. Although toys are, in general, targeted towards children, the majority of toys are also aimed at parents. For the purpose of this paper, we will focus on toy marketing towards children.

To understand the basic characteristics of toys, we need to understand their evolution. The history of toys is as old as the history of human civilization. The earliest evidence of toys is found in archeological sites (Mann 1975). These toys were largely made from natural materials such as clay, sticks, and rocks. It is likely that prehistoric families had toys of animals, toy weapons to play-hunt these animals, and representations of mythical characters. These toys not only helped children en-

---

[1] As defined by Dictionary.com at http://dictionary.reference.com/browse/toy

tertain themselves, they no doubt helped them to develop skills and to role-play adult activities, while emphasising/reinforcing traditional gender roles. Successive excavations reveal that children played with dolls and balls in ancient Roman times.

It should be noted that many children's toys did not change much in terms of form over the centuries. As such, the modern toy industry emerged after the Industrial Revolution with the introduction of new manufacturing technology. For example: dolls were used as toys in many ancient societies. But dolls have evolved as technology evolved over many centuries, especially after the rapid industrialization that allowed toys to be mass-produced in factories at low cost (Brown 1990). In the last 20 years, technology has led to development of toys with new forms and connectivity. Toys such as electronic and computer games are increasingly popular among children. In recent years we have also seen evolution of computing devices as toys. For example: smart phones such as the early BlackBerry were utilitarian devices for adults to communicate, but with the introduction of the iPhone, smart phones suddenly become a device to improve productivity *and* a toy for both adults and children. Subsequently, tablets became the next computing product to cross over from a product for productivity to a toy. These cross-overs from utilitarian information technology products to toys have created immense opportunities for information technology firms and toy manufacturers to innovate and collaborate to change the landscape of toy markets. We have already seen examples of such combinations with the introduction by traditional toy manufacturers of LeapPad and Innotab tablets for children. In fact, such tablets have become the number-one gift to children in the UK and in North America as other toys struggle to generate growth in sales (London 2012). In this paper, our focus will be on MCTs that are not tablets. Next we outline the current state of the global toy market.

Current State of the Toy Market

This section provides an overview of the current state of the toy market. Globally, the toy industry is significant with annual sales of US$ 84 billion in 2012 (Euromonitor International 2014). The growth in toys is largely influenced by both the demand and the supply sides of the market. On the demand side, a growing population and rising incomes in emerging markets offer tremendous market opportunities for growth for both traditional toys and MCTs. On the supply side, the development of new products and e-commerce have generated increased sales and revenues for toys. This section is divided into three parts: (1) toy market size, per capita spending, and growth rate; (2) toy categories; and (3) major toy companies.

Toy Market Size, Per Capita Spending, and Growth Rate

On a global basis, the toy market is currently experiencing slow growth. Table 1 presents the toy market size, per capita spending, and growth rates by region in 2012. As seen in Table 1, North America is the largest toy market, followed by

Table 1 Toy market size, per capita spending, and growth rate by region in 2012. (Sources: Euromonitor International, Passport, 2014)

Region	Market size (in billions of US dollars)	Per capita (in US dollars)	Growth rate 2011-12 (%)
Asia Pacific	20.36	5.2	4.0
Australasia	1.67	61.2	0.4
Eastern Europe	5.39	16.3	7.1
Latin America	10.43	17.4	11.0
Middle East & Africa	4.86	3.8	6.6
North America	21.52	61.7	−0.5
Western Europe	19.96	40.8	−2.0
World	84.19	12.0	2.4

The world total is slightly greater than the sum of all regions due to rounding

Asia Pacific and Western Europe. It should be noted that the strongest growth rates were in Latin America (11.0%), Eastern Europe (7.1%), the Middle East and Africa (6.6%), and Asia Pacific (4.0%), while the toy markets in both North America and Western Europe remain stagnant with negative growth rates. In essence, global growth has been driven by emerging markets (e.g. India, China, and Russia) as the sales of toys continue to decline in developed economies. Based on the current forecast, the Asia Pacific region is expected to surpass North America as the largest toy market by 2014. In 2012, China became the second-largest toy market after the US (Euromonitor International 2014). With an impressive growth of 11%, China was the most rapidly growing toy market during 2012. In Eastern Europe, Russia accounted for almost 60% of the toy market. Thus, the emerging markets of China, India, and Russia offer tremendous growth opportunities for MCT products.

Toy Categories

The toy market is highly gender-segmented (Auster and Mansbach 2012). Upon entering a toy store, one sees immediately which toys are targeted to girls versus boys based on the color scales: pink shades for girls and a mix of green and browns for boys (Sparrman 2009). Table 2 presents a summary of selected toy categories by

Table 2 Selected toy categories. (Sources: Based on information from several toy retailers)

Category	Boys	Girls
Action figures	Toy soldiers, Star Wars	Disney Princess, Furry Boom
Arts and crafts	Doorknob hanger, crayons	Paint 'n' sparkle, play-doh
Construction toys	Legos, blocks	Toy kitchen, toy store
Dolls and accessories	King Kong doll, Buzz Lightyear	Hello Kitty, barbie doll
Dress-up clothing	Superhero, cowboys	Ballerina, princess, barbie
Games and puzzles	Trivial pursuit, battleship	Rummikub, Candy Land
Model vehicles	Toy cars, fire trucks, planes	Barbie bicycle
Sports	Ice hockey, basketball hoop	Figure skates

gender. Boys' toys include action figures, soldiers, Star Wars, Legos, superheroes, cowboys, cars, fire trucks, and planes. In contrast, girls' toys include toy kitchens, Hello Kitty, Barbie dolls, princess items, and paint 'n' sparkle. The amount of toys a child owns has increased from a few toys during the first half of the twentieth century to hundreds of toys by the end of the century (Sandberg and Vuorinen 2008).

Major Toy Companies

The global ranking of toy companies is presented in Table 3. With sales of more than US$ 10 billion, Mattel is the largest toy company with the highest market share of 12.4 % in 2012. Several companies (Bandai Namco, Hallmark Cards, VTech, and MGA Entertainment) were able to maintain their relative market shares between 2008 and 2012. The performance of Lego is impressive, as the company increased its market share from 3.6 % in 2008 to 6.3 % in 2012 due to the success of its construction sets for each age group. All these major toy makers already have MCT products. Moving forward, the biggest challenge for them will be the integration of mobile capabilities in their traditional toys without sacrificing the current level of satisfaction among users.

It should be noted that this is not the first time toy makers have evolved and incorporated new technologies in order to keep products relevant to consumers. As such, the toy market is a dynamic one with constant changes due to globalization and technological advances. On the consumer side, we have witnessed a convergence of consumer tastes and preferences for toys in different countries. When a new toy appears in the North American market, consumers in Asia and Latin America demand the same product in their own countries. On the production side, a number of toy companies have shifted their manufacturing activities to Asian countries such as China and Thailand in order to take advantage of lower costs.

Next we discuss basic characteristics of toys to formulate insights into how information technologies can be successfully incorporated into traditional toys.

Table 3 Global toy shares by toy companies (in percentage; Sources: Euromonitor International, Passport, 2014)

Company	2008	2010	2012
Mattel Inc.	12.3	12.1	12.4
Hasbro Inc.	8.3	8.1	7.6
Lego Group	3.6	4.9	6.3
Takara Tomy Co. Ltd.	2.8	2.9	2.6
Bandai Namco Group	2.2	2.4	2.3
Hallmark Cards Inc.	1.4	1.6	1.6
VTech Holdings Ltd.	1.1	1.2	1.3
MGA Entertainment Inc.	1.3	1.3	1.2
Spin Master Ltd.	1.0	1.5	1.1
LeapFrog Enterprises Ltd.	0.9	0.9	1.1

Characteristics of Toys

Toys play an important role in the development of children (Auster and Mansbach 2012). Although toys have changed over the years, the basic concept remains the same. Toys are used to stimulate pretend play, the development of cognitive skills, and social play with other children (Blakemore and Centers 2005). It should be noted that the distinction between a toy and a utilitarian object is subtle. In the case of a smart phone, parents can use it as a device to improve their work productivity and the children can use it as a toy. Hence, a device can be both a toy and an object to perform certain tasks. From our perspective, a device becomes a toy when it generates pleasure in the process of its utilization. To provide insights into the toy development process, we need to understand the basic characteristics of toys per se. As such, the characteristics of toys play a significant role in the marketing of these products to consumers (Auster and Mansbach 2012). Goldstein et al. (2005) discussed extensively various characteristics of toys and games. Based on their summary we highlight three basic characteristics of successful toys:

1. Scaled Representations: Almost all successful toys are miniature representations of characters or tools. These characters and devices can be either real or mythical. For boys, toy guns are a classic example of scaled representation of weapons. On the other hand, for girls, dolls are a great example of miniature humans. MCT provides an excellent opportunity for toy makers to introduce scaled representations of new products.
2. Repository of Imagination: Almost all successful toys provide players the opportunity to construct scenarios using imagination to immerse themselves in specific role play. This is the reason why toys based on *Star Wars* and other movies are extremely popular. MCT with their potential multi-media capabilities can easily enhance users' ability to imagine and create scenarios.
3. The Possibility of Creative Destruction: Almost all successful toys are malleable. Players can break them apart or mold them based on new, creative ideas. The most enduring and successful toys become timeless because of this property. A classic example of this property is embedded in the most successful toy company of the last century, Lego. Lego allows players to build objects in infinite combinations, play with the constructed object, and later build a new one. By sticking to this basic property of toys, Lego has become one of the most enduring toy companies of all time. Given the complex nature of integration of hardware, software, and traditional toy form, existing MCTs have not been able to develop products that can be creatively reconstructed and played with in many different forms. We believe this is one of the biggest challenges facing makers of MCTs.

For MCT to be successful, toy makers need to focus on these three properties. The challenge is developing a highly complex toy that can embody these three elements while also providing significant enhancements to the user experience. The challenge arises from the fact that MCTs are a marriage of software, hardware, and traditional toy concepts.

Another big challenge for MCTs will be breaking the gender barriers. The challenge in this context is twofold. First, it is generally argued that there exist gender differences in how information technology is perceived and used as well as in the actual usage rate (Jackson et al. 2008). Second, toys are historically gender-specific. In general, boys' toys are more violent, competitive, exciting, and dangerous compared to girls' toys, which are more physically attractive and associated with nurturance and domestic skills (Fisher-Thompson et al. 1995; Blakemore and Centers 2005). However, over the years there is evidence of girls wanting toys typically classified as "toys for boys" and vice versa (Marcon and Freeman 1996). Until now, video games and similar toys have been more successful with male than female consumers. For MCT to achieve its full potential, it needs to figure out how to make these toys more accessible to female consumers.

Toys and Marketing Research

Most studies on toys have largely focused on historical, educational, and psychological issues. The four major research areas are: the history of the toy industry (Brown 1990, 1993; Cross and Smits 2005), toys by gender (Fisher-Thompson et al. 1995; Blakemore and Centers 2005; Auster and Mansbach 2012), child development (Hornik et al. 1987; Sandberg and Vuorinen 2008), and toy safety (Taylor et al. 1997). There are only a few studies on toys in the business literature. Several studies have focused on the supply chain (Law and Chan 2003; Egels-Zanden 2007; Chan and Chin 2007) and toy recalls (Beamish and Bapuji 2008; Teagarden 2009).

In an attempt to understand the toy market, Hogan (2007) developed a conceptual framework to explain how trust is created between toy companies and parents. Lin (2010) examined the relationship between consumer personality traits and brand loyalty involving toy purchases. In a recent study, Gardner et al. (2012) examined the role of advertising and parents' perceptions of advertised toys.

The lack of research on marketing of toys is somewhat surprising given the large size of the toy industry. The next section discusses challenges and marketing strategies for MCTs using key strategic concepts of marketing management.

Challenges and Marketing Strategies for MCT

Based on the basic definition of MCTs, this category of toys has the potential to be highly successful in the market. Conceptually, these toys are a marriage between two very popular product categories with children: software and related computing hardware, and traditional toys. On the computing side, according to Common Sense Media, 38 % of children under the age of 2 used a mobile device for playing games, watching videos, or other media-related purposes in 2013. In 2011, that number was only 10%. And by the age of 8, 72% of children have used a smart phone, tablet, or similar device. In terms of growth, in 2013 children in the 0–8 age group spent an average of 15 min a day using mobile devices; that's up from 5 min a day in 2011 (Common Sense Media Press Release 2013). In fact, tablets and other mobile devices have become so popular with children that they, in fact, are competing for the share of spending for children's gifts. In 2013, tablet computers designed specifically for kids were the second most popular gift in the UK.

On the other hand, traditional toys, although still popular, either have stagnant sales (especially in the US market) or significantly slower growth (in the global market). Hence, the traditional toy industry could gain a competitive advantage from this marriage with technology rather than the other way around. And the success and failure of this, like in any marriage, will depend on whether the result (mobile computing + traditional toy) brings an improved toy-user experience than if the two remain separate. In the case of any new MCT, the experience of the users has to be better than the experience they could get by using each of the components separately. This has been a challenge in many other industries. Before Apple came up with the iPod, the hardware, software, and music industries struggled to develop a product that enhanced the value of all three components. So, if a company is developing a toy with mobile computing components, then the experience of the users with the new MCT has to be better than if the users were just using the toy itself. To present our insights regarding how to overcome this challenge, we will focus on the "4Ps" of marketing strategies: product, price, promotion, and place.

Product We have not yet seen a blockbuster product in this category from the traditional toy manufacturers. Every major toy manufacturer is involved in developing and introducing products in this category. Launched in 2012, *Apptivity* by the Mattel line of toys is a classic example.[2] In this case, Mattel is targeting mobile device owners (tablets or phones) by developing apps that can enhance the gaming from a traditionally designed toy (such as a figurine). Similarly, the Hong Kong-based toy company *Apptoyz* has been in the business of enhancing traditional toys that can interact with mobile device-based apps.[3] In both these cases, they are relying on the popularity of Apple's mobile devices. Certainly this approach in developing MCT has distinct disadvantages:

1. Products cannot be mass-marketed, as only kids with access to mobile devices will be able to play with these toys. Note that even device-owning parents may not be interested in buying these toys for their children because they may/will have to share their mobile devices. In terms of accessibility, as more and more consumers acquire smart phones and tablets, the challenge of mass marketing will potentially be resolved.
2. Relying on another party to build or enhance your product creates tremendous opportunities for hardware or software makers to get into this business themselves if these toys become highly profitable.

---

[2] For further details, please refer to: http://www.mattelapptivity.com/
[3] For further details please refer to: http://www.apptoyz.com/

3. In terms of ease of use, these toys add another layer of complexity or cost of use. For a user, this cost has to be lower than the enhanced benefits from the MCTs. Otherwise success will always elude MCT makers.

In the toy industry, licenses are important for classic toys such as Star Wars, Spiderman, Disney Princess, Hello Kitty, Barbie dolls, and so on. New toys cannot be successful without the protection of global licenses. It should be noted that toy manufacturers can develop toys with mobile computing abilities built into them. But such a strategy will certainly increase the cost and thereby price of these products significantly. One approach that can significantly lower the cost of such a strategy is to first develop a mobile device that can work with multiple games. But again, such a strategy can be risky and expensive to develop. For this reason, we believe that in the near future, most MCTs will rely on this marriage of convenience between mobile devices and traditional toy forms. It is expected that new product development will be centered on smart phones and children's tablets involving some classic toys such as Star Wars, Spiderman, Hello Kitty, and Barbie dolls.

Promotion Since these toys are part of a completely new product category, marketing will play a key role in the successes and failures of these products. Potential customers need to be convinced of the benefits of these toys. Most households with children already own toys and, at the same time, own multiple mobile devices. Children are already using the toys and, in some cases, mobile devices separately for their enjoyment. The challenge for the marketers is to convince children that a combination of these two products can be more enjoyable. Similarly, promotional campaigns need to convince parents to allow children without access to such mobile devices to either own or borrow the devices to play with these toys. The value proposition to the children is easier to make, as these toys are novelties for them. On the other hand, the value proposition to the parents needs to go beyond the argument of enhanced enjoyment for the children. Since the emerging economies are the fastest-growing regions for toys, toy companies should focus their marketing and advertising for MCT products in these geographical areas.

Place Because MCTs are a completely new product category within the toy industry, toy companies need to think in terms of revamping the marketing channels to sell these items. They need to be delivered using full-service retailers or any other channel where full-scale demonstration can be implemented. Apple's strategy of setting up its own store to sell its category-defining new products can certainly be used as a template to sell new toys. There is a growing trend of shopping for toys in grocery retailers and drug stores in developed countries. However, the importance of traditional toy stores (e.g. Toys "R" Us) should not be overlooked.

Price In the case of pricing, toy companies can certainly innovate and forgo standard pricing strategies of traditional toys. First, given that these toys will need two components, toy and app(s), toy makers can price these two components separately. Second, toy companies can devise strategies to release enhancements of apps over time. Such a strategy will help toy manufacturers price these enhancements separately, thereby creating a future revenue stream from the sale of a single toy. Third, toy companies can also introduce bundle pricing for the mobile computing toys along with small action figures or vehicles.

In terms of segmentation given the gender differences in toy preference, makers of MCTs need to make sure that toy development incorporate these distinctions and adequately considers how male and female consumers interact with toys. Given that the video game industry is well known for its lack of market reach to female consumers (Hartmann and Klimmt 2006), developing toys that will appeal to girls will take some thought because mobile computing toys critically rely on similar interfaces and technology as the video game industry.

Concluding Remarks

The toy market is currently at crossroads. With the emergence of MCTs, the rate of innovation in toys has increased significantly, creating a sudden potential for information technology firms to enter toy markets on their own or develop joint ventures with traditional toy manufacturers. Hence, MCTs hold immense economic promise not only for traditional toy makers, but also for mobile software and hardware companies. However, this promise has yet to be realized. One reason is that there is a lack of an integrated product that would function such as the Apple iPhone in the mobile phone category. Another possible explanation is that the products are not generating significant new benefits for consumers.

Toy companies and the mobile computing industries need to think beyond just the novelty factor to create enduring value propositions for consumers of MCT products as opposed to offering only niche products or fads. The market players also need to carefully consider the value propositions for the parents and not only the child users. It is hoped that this paper provides broad foundational guidelines from a marketing perspective on how to develop and sell MCT products.

References

Auster, C. J. and Mansbach, C. S. 2012. The gender marketing of toys: an analysis of color and type of toys on the Disney store website. *Sex Roles*, 67(7/8): 375–388.

Beamish, P. W. and Bapuji, H. 2008. Toy recalls and China: emotion vs. evidence. *Management and Organization Review*, 4(2): 197–209.

Blakemore, J. E. O. and Centers, R. E. 2005. Characteristics of boys' and girls' toys. *Sex Roles*, 53(9/10): 619–633.

Brown, K. D. 1990. The children's toy industry in nineteenth-century Britain. *Business History*, 32(2): 180–197.

Brown, K. D. 1993. The collapse of the British toy industry, 1979–1984. *Economic History Review*, 46(3): 592–606.

Chan, T. C. T. and Chin, K. S. 2007. Key success factors of strategic sourcing: an empirical study of the Hong Kong toy industry. *Industrial Management and Data Systems*, 107(9): 1391–1416.

Common Sense Media Press Release. 2013. New Research from Common Sense Media Reveals Mobile Media Use Among Young Children Has Tripled in Two Years. Available at: https://www.commonsensemedia.org/zero-to-eight-2013-infographic

Cross, G. and Smits, G. 2005. Japan, U.S. and the globalization of children's consumer culture. *Journal of Social History*, 38(4): 873–890.

Egels-Zanden, N. 2007. Suppliers' compliance with MNC's codes of conduct: behind the scenes at Chinese toy suppliers. *Journal of Business Ethics*, 75: 45–62.

Euromonitor International. 2014. *Passport*. London, UK.

Fisher-Thompson, D., Sausa, A. D., and Wright, T. E. 1995. Toy selection for children: personality and toy request influences. *Sex Roles*, 33(3/4): 239–255.

Gardner, M. P., Golinkoff, R. M., Hirsh-Pasek, K., and Heiney-Gonzalez, D. 2012. Marketing toys without playing around. *Young Consumers*, 13(4): 381–391.

Goldstein, J., Buckingham, D., and Brougere, G. 2005. *Toys, Games, and Media*. Routledge.

Hartmann, T. and Klimmt, C. 2006. Gender and computer games: exploring females' dislikes. *Journal of Computer-Mediated Communication*, 11: 910–931.

Hogan, S. P. 2007. Creating parental trust in the children's toy market. *Young Consumers*, 8(3): 163–171.

Hornik, R., Risenhoover, N., and Gunnar, M. 1987. The effects of maternal positive, neutral, and negative affective communications on infant responses to new toys. *Child Development*, 58(4): 937–944.

Jackson, L. A., Zhao, Y., A. Kolenic III, Fitzerald, H.A., R. Harold, and Eye, A.V. (2008), Race, Gender, and Information Technology Use: The New Digital Divide. *Cyber Psychology and Behavior* 11(4): 437–442.

Law, C. K. and Chan, S. F. 2003. Panorama of toy design and development in Hong Kong. *Journal of Materials Processing Technology*, 138: 270–276.

Lin, L. Y. 2010. The relationship of consumer personality trait, brand personality and brand loyalty: an empirical study of toys and video game buyers. *Journal of Product and Brand Management*, 19(1): 4–17.

London, B. 2012. Must-have Christmas present for three-year-olds? The LeapPad2 … Tech toys and children's tablets top festive gift list. *The Daily Mail*. Available at: http://www.dailymail.co.uk/femail/article-2208340/Tis-season-tech-toy-kiddie-tablets-Toy-Insider-gift-list.html

Mann, T. 1975. How toys began. *Proceedings of the Royal Society of Medicine* 68(1): 39–42.

Marcon, R. A. and Freeman, G. 1996. Linking gender-related toy preferences to social structure: changes in children's letters to Santa since 1978. *Journal of Psychological Practice*, 2: 1–10.

Sandberg, A. and Vuorinen, T. 2008. Dimension of childhood play and toys. *Asia-Pacific Journal of Teacher Education*, 36(2): 135–146.

Sparrman, A. 2009. Ambiguities and paradoxes in children's talk about marketing breakfast cereals with toys. *Young Consumers*, 10(4): 297–313.

Taylor, S. I., Morris, V. G., and Rogers, C. S. 1997. Toy safety and selection. *Early Childhood Education Journal*, 24(4): 235–238.

Teagarden, M. B. 2009. Learning from toys: reflections on the 2007 recall crisis. *Thunderbird International Business Review*, 51(1): 5–15.

Toys and Mobile Applications: Current Trends and Related Privacy Issues

Gary Ng, Michael Chow and André de Lima Salgado

Abstract A toy is a product that is intended for play. Common examples of toys include dolls, board games and video games. Based on the reports from the 110th Annual American International Toy Fair, toy manufacturers are increasingly incorporating mobile applications into their products through smartphones. This technological advancement gives users the opportunity to augment the physical world as they play through a user interface of information services. These toys capture the user's physical activity state (e.g., walking, standing, running, etc.) and store personalized information (e.g., location, activity pattern, etc.) through the mobile device. By adding a digital component onto an analog device, a greater potential for a more personalized and dynamic play experience, such as augmented reality, is unlocked. In addition, this play experience can be further modified after point of sale through a developer's content or update release through their digital service. As smartphones and mobile applications play a larger role in the everyday lives of their users, toy manufacturers commonly face the challenge of better understanding the modern consumer's needs as well as their concerns. Issues such as privacy and data encryption, which were not common problems in the toy industry, are suddenly becoming necessary due to these trends. This book chapter also discusses a case study of toy guns in Brazil.

Keywords Mobile computing · Toy industry · Human-computer interaction · Interactive systems · Augmented reality

G. Ng (✉)
Faculty of Business and Information Technology, University of Ontario Institute of Technology, Oshawa, Canada
e-mail: ng.gary.kc@gmail.com

M. Chow
Faculty of Business and Information Technology, University of Ontario Institute of Technology, Oshawa, Canada
e-mail: me@michaelchow.ca

A. L. Salgado
Computer Science Department, ICMC-University of São Paulo, São Paulo, Brazil
e-mail: andrelima.salgado@gmail.com

© Springer International Publishing Switzerland 2015
P. C. K. Hung (ed.), *Mobile Services for Toy Computing,* International Series on Computer Entertainment and Media Technology, DOI 10.1007/978-3-319-21323-1_4

Fig. 1 Ancient Egyptian
toys. (Maspero 1895)

Section 1. Introduction

A Brief Background of Toys

Throughout history, toys have been a vital tool to educate and entertain human beings. Evidence shows that toys have existed as far back as the time of the ancient Egyptians where simple toys were created with clay or wood (Maspero 1895). Figure 1 depicts an illustration of what some of those era's toys looked like. However, despite being a part of human society for thousands of years, toys as a commercial product did not exist until relatively recently. The Child Study Movement, that took place around 1900, led to an increase in child development (Burton 1997), which inevitably led to a larger focus to different facets of a child's life, including toys. The changing states of human attitudes and the continuous development of technology kick-started the modern toy industry and allowed it to thrive.

As technology developed, toymakers came up with new and innovative ways to leverage new technology to make their toys better and more desirable (Toy Industry Association 2013). A larger focus was placed on the design of the toys and how they can be used for play and education. It was no longer about making a generic toy but rather the right toy for the targeted consumer. In countries such as Canada, Brazil and Japan, toys became more and more complex; gone are the days where people played with simple wooden figures painstakingly crafted from a block of wood. Modern toys now take on many different forms, ranging from odd looking, malleable objects to intricate models made up of tiny plastic and metal parts.

Despite the advancements in toy technologies, the potential of traditional toys are still very limited because they are often static items that do not change. It is up to the player's imagination to bring the toy to life and to give the play experience meaning, but if the toy does not change then eventually the player's attention will be lost. For example, an action figure can have its limbs positioned in many possible ways. Those toys may even have different accessories or clothing that can modify

Fig. 2 Kamen Rider
Ichigo Henshin Belt
released in 2008.
(Bandai 2014b)

Fig. 3 Kamen Rider
OOO Henshin Belt
released in 2013. (Ban-
dai 2014a)

the toy. However, the functionality of that toy is still very limited. If the toy's content never changes, the player will likely become bored of that toy faster than if it did change because the play experience remains the same.

To account for the fact that traditional toys are static play experiences, some toys are designed with simple digital elements to extend the lifecycle of the toy. For example, Bandai has been releasing replica toy belts from the popular Japanese TV series *Kamen Rider* for many years. Early versions of those belts such as the one for *Kamen Rider Ichigo*, which is shown in Fig. 2, featured a simple toy belt that emitted basic sounds and lights (Bandai 2014). More recent belts such as the one for *Kamen Rider OOO*, as shown in Fig. 3, are equipped with a simple digital sensor that will allow the toy to process basic inputs from the player and output a range of dynamic results (Bandai 2014). Toy medals that come with the belt have chips inside of them, which can be scanned by a sensor, which will emit sounds and lights based on that specific medal. This allowed Bandai to continually release new medals that could be used with the belt and as a result made the toy more dynamic than a traditional toy.

The popularity and power of digital technology influenced toy developers to create toys that integrate many more digital aspects. Some toys like the *Tamagotchi*, a pet simulator released in the 1990s, are almost entirely digital experiences as the physical toy merely serves as an interface to the digital game. The player interacts with a *Tamagotchi* by feeding and playing with the virtual pet housed in the egg-shaped casing shown in Fig. 4. Throughout the pet's lifecycle, it grows and thus enables the player to interact with it in new ways. This type of toy is much more dynamic because the digital elements allow the toy to change. Despite being an egg shaped peripheral, the gameplay is entirely digital and as a result there are many more possibilities for the user as he or she plays with the toy.

A Brief Background on Digital Technology

The advent of personal computers and digital technology gave rise to a new form of entertainment: video games. Unlike traditional toys, video games are purely digital

Fig. 4 A screenshot of
Tamagotchi L.i.f.e., a
Tamagotchi simulator.
(Bandai Namco Games
Inc. 2013)

products and thus are not limited to physical constraints. This results in a highly
flexible and dynamic play experience that changes based on the player's input. The
first commercial video game, *Pong*, released in 1972, is a simple two-dimensional
game where two paddles can be controlled to hit a ball back in forth. In the four
decades since it was released, video games have grown exponentially in complexity and popularity which spawned many generations of consoles, as illustrated in
Fig. 5. Today, it is a multi-billion dollar industry that continues to expand and develop at a rapid pace (Nayak 2013).

One of the aspects of why video games have been successful is because they are
arguably the most interactive mediums that exist. This allows the players to fully
engross themselves in the game experience, unlike when playing with toys or reading a book where they must use their imagination to immerse themselves. Video
game immersion has also shown to be an effective way to teach and train players for
a variety of skills. This is explored in a field known as serious gaming where games
are designed to go beyond the pure entertainment aspect and are used to affect the
player in a variety of ways (Bellotti et al. 2013).

Digital technology is advancing at an incredible rate, which is prompting the
development of powerful devices and software that was unthinkable a mere decade
ago. Moore's law is an observation made in 1965 that stated that the number of tran-
sistors on integrated circuits will double approximately every 2 years (Moore 1965).
This observation has been relatively accurate throughout the last few decades, and
as a result there has been an exponential growth in computing power over the years.

The increase in computing power also made it possible to downsize computers to the point where every day consumers can own a powerful computer that is
small enough to fit into their pockets. In 1969, a guidance computer with 64 KB of

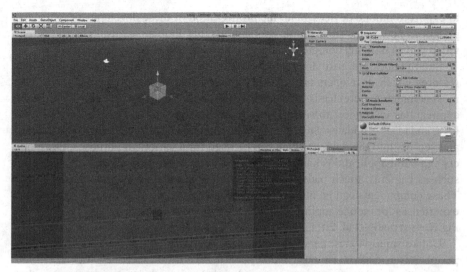

Fig. 5 A screenshot of a simple mobile game being developed with *Unity*. (Nokia Developer 2014)

memory and 0.043 MHz of processing power successfully controlled a spaceship and landed the first humans on the Moon (Fightglobal 2009). In comparison, one of the *Samsung Galaxy S4* smartphone models released in 2013 contains a 1.9 GHz quad-core processor with up to 64 GB of hard drive memory and an additional 2 GB of random access memory (Samsung 2013). An example of how far technology has come can be seen by comparing the computing power of the Apollo 11 Guidance Computer to a modern smartphone.

The growing popularity of smartphones and tablets has also seen an increase of mobile applications and games. According to the International Data Corporation (IDC), hundreds of millions of smartphone devices were shipped worldwide in 2013, with over 90 % of those running the Android or iOS operating systems (International Data Corporation 2013). The interest in game applications is so high that, among user experience research, this is one of the fields that contains more specific heuristics for evaluating usability of game apps (de Lima Salgado and Freire 2014).

Since the overwhelming majority of smartphone users use Android or iOS, it makes it easier for developers to connect with their target market since they only need to support two platforms to reach most of the users. But the amount of investment that Microsoft is doing on this area is making the Windows Phone more successful. On amazon.com (Amazon 2014), one of the most sold smartphone devices is a popular model that runs the Windows Phone operating system.

The massive popularity of these devices combined with the sheer processing power available allowed developers to create millions of novel applications for a variety of purposes. Many pieces of software and tools have been created to facilitate the development of these applications. For example, Unity (see Fig. 5) is a tool for mobile games development, it is an easy-to-use game engine, which has features to export to mobile devices. With the proliferation of these development tools, applications now range from simple applications that simulate a flashlight, to full 3D video games utilizing complex shader graphics and network capabilities.

Smartphones are also jam packed with a variety of different components and features which developers can take advantage of when making applications. The most common feature that is used is wireless connection to the Internet. This allows applications to connect to servers online, do a variety of tasks such as send data logs, and receive updates. Smartphones also integrate a variety of other components and sensors into the device, which can be used for a variety of purposes. Developers can build applications knowing that their target platform will have components such as cameras, touch pads, accelerometers, microphones, vibrators, and more. These components can be utilized in the development of innovated interfaces and experiences for mobile applications that can engage the user in unique and interesting ways (Huang et al. 2011).

Unlike smartphones, personal computers typically do not have any additional peripherals aside from a mouse and keyboard. As a result, most developers will not create software that uses any other peripherals because they cannot guarantee that their user will have access to it. On the other hand, lists of included components and sensors can be easily found for every smartphone. As a result, developers can create mobile applications specifically targeted to certain systems and design their product in a way that delivers unique experiences that is not often found on traditional computer platforms. One popular form of this is augmented reality (AR).

Augmented Reality

According to Merriam Webster (2014a), augmented reality is "an enhanced version of reality created by the use of technology to overlay digital information on an image of something being viewed through a device". This essentially means that digital elements are overlaid onto the real world when viewed through a virtual interface such as a screen on a smartphone. Typically, reality is captured through a camera and then additional elements are added through software to augment it. Examples of this are additional pieces of information displayed in a heads-up display (HUD) or to render digital avatars and characters into the scene itself. Since digital components are downsizing and become commonplace, more and more augmented reality based applications are hitting the market.

A popular example of an AR enabled device is the Nintendo 3DS handheld video game console. Every Nintendo 3DS system comes with AR cards which can be recognized by the built in software and display 3D models overlaid on the device's screen (Nintendo 2014). An example of this is shown in Fig. 6. This is possible because the software is able to recognize specific patterns in the video feed from the device's camera and when something is matched, the appropriate actions are executed. Many games on the Nintendo 3DS also leverage this technology to add a bit of AR to their gameplay. This technology is also available on mobile devices

Fig. 6 An example of the Nintendo 3DS AR cards. (Nintendo 2014)

such as smartphones and tablets as modern mobile devices all come with cameras and hardware that can support this type of processing. As a result, it is possible for application developers to develop software that provides the user with a full spectrum of AR experiences.

Section 2. Current Trends in the Toy Industry

Toys with Digital Components

A recent trend that has appeared is the proliferation of analog toys with digital components. At Toy Fair 2014 in New York City, a variety of technologically augmented toys were showcased to engage and educate children (Bradford 2014). These types of toys are more interactive and accommodates for the user's learning level and aptitude to achieve the optimal learning and entertainment experience.

One way that toymakers have been adding digital elements to toys has been through mobile applications installed on smartphones. The toy Sphero (Google Play[a] 2014), a robotic ball that is controlled by the player's smartphone, is a product of this trend. The player downloads and installs a special mobile application onto their phone and links it with their toy. Once that is completed, the player can control the robot's movements using the application's interface.

Sphero is an example of a type of remote controlled toy that has embraced the power of mobile applications. The mobile application shown in Fig. 7 can be used as a controller to move the toy around, similarly to other remote controlled toys. However, since Sphero was designed to work with a mobile device, which has the

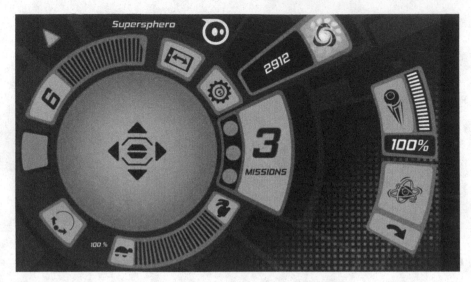

Fig. 7 The mobile application interface for Sphero. (Google Play[a] 2014)

potential for much more dynamic experiences, there is a lot more that can be done than just move the toy around.

For example, the application has missions where the player is tasked with completing certain objectives such as crashing into five things within 25 sec or maintaining a certain speed for a certain amount of time. The application also keeps track of the player's progress through levels and other types of rewards. The device's various capabilities are also used by the application to perform tasks such as allowing the player to control the ball by either sensing gestures on the touch screen or by utilizing the gyroscope to measure the orientation of the device. This type of content is unique to mobile applications, and it shows how an existing type of toy can be augmented by utilizing the potential found within mobile devices.

Another example of a toy that utilizes the smartphone's functionality is Toymail, a small battery powered toy depicted in Fig. 8. By speaking into the microphone of their smartphone, users can send messages to their child's toy where their message is relayed in a cartoon voice (Brian 2014). The child can then send a recording of their own back to the sender of the message. The manufacturers of this toy understood that there is a problem of getting children to talk on the phone, and solved it by introducing digital elements to create a product that is more engaging to children.

Similar to Toymail is the Brazilian toy named Elo (Fundação Hospital Amaral Carvalho 2014). Elo was developed by the Amaral Carvalho Foundation, a Brazilian hospital located in Jaú, inside São Paulo state. As shown in Fig. 9, kids undergoing cancer treatment can receive audio messages by pressing Elo's hand. A specific hospital employee is responsible for filtering the messages and forwarding them to each toy. This innovative toy is showing a new way of supporting children who are fighting against cancer and is not commercialized yet.

Fig. 8 A child with a toymail. (Brian 2014)

Fig. 9 Elo toy, for kids under cancer treatment. (Fundação Hospital Amaral Carvalho 2014)

Toys with Integrated Smartphones

Smartphones can also be physically integrated into the toy to further expand on the user's play experience. Popular examples of this are shown through toy guns that have mounts where smartphones can be attached. Once the smartphone is mounted onto the gun, a mobile application can be run on the phone to augment the player's play experience with the toy. Oftentimes, the application displays a reticle and acts as an interface to an AR world. Toy guns that use this type of technology, such as the Tek Recon (Tek Recon 2014) blaster, often fire small plastic bullets and are reminis-

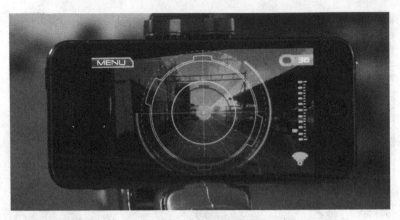

Fig. 10 The Tek Recon application with radar mode enabled. (Tek Recon 2014)

cent of traditional toy guns. They function similar to a traditional toy gun, allowing players to leverage their existing knowhow to play with this type of toy. Once a smartphone is secured onto the mount on the gun, the toy's mobile application can be enabled, allowing for an enhanced and personalized experience.

The use of a smartphone enables many new features that would not be possible without the gun. For example, a team mode can be started where players with the guns and mobile application can join teams and engage in a team battle. Through the smartphone's microphone and speakers, the player can communicate with other members on their team to plan strategies and execute tactics. The smartphone also displays an on-screen radar that shows the location of their team members and objectives. Fig. 10 depicts an image of the Tek Recon blaster and its mobile application acting as a heads-up display (HUD). This is akin to what would be used in traditional first-person shooter (FPS) video games. Games such as *Metro 2033* by THQ feature completely diegetic interfaces, which are used to fully immerse the player in the game world (Stonehouse 2014). However, players still view the game through a TV screen, which breaks the immersion factor.

The Tek Recon blaster, however, provides players the opportunity to bring their gaming experience to real life by mounting a smartphone on their analog gun to provide the player with "an interface complete with interactive heads up display, live chat, and radar tracking" (Tek Recon 2014). This enables the user to access information that normally would not be available to them, resulting in a more immersive and filling gameplay experience.

Integration of Digital Components to Analog Toys

The Tek Recon gun functions as a traditional toy gun without the mobile application. Other toys such as Appfinity's AppBlaster, shown in Fig. 11, offer a similar play experience, but they forgo the ability to physically fire bullets and entirely

Fig. 11 The AppBlaster v2 showcasing the smartphone mount. (Apptoyz 2014)

focus on the digital gaming aspects. Like the Tek Recon gun, the AppBlaster is a toy gun that has a mount where a smartphone can be secured to. However, the applications that are available for purchase or download are essentially traditional mobile games, which use the toy gun as a physical interface to the game.

Many of the AppBlaster applications are games where the player shoots digital targets. An example of this is the free application *Alien Attack*, which when loaded on a smartphone utilizes the phone's camera to display a feed of the player's environment. Then, cartoon aliens spawn off-screen and head toward the player. The player must then shoot the aliens by aiming the gun in that direction and pulling the trigger. The application uses the phone's orientation to calculate where the player is facing and any taps on the screen represent a bullet being shot. This means that the game could be played without the gun as the player could just move the smartphone around and tap the screen.

Both of these two types of guns use similar technologies in their products; yet, the actual play experience is different due to the functionality of their corresponding mobile applications. For example, the Tek Recon application augments the play experience provided by the toy gun. The focus is still on the gun's capabilities of shooting plastic pellets. The mobile application merely enhances the activity by providing the user with a window into an augmented reality where they have access to more information about the game state. The AppBlaster in contrast, focuses on the mobile application as the main activity. This means that the toy gun is not as important and just serves as a novelty interface to a mobile game.

Table 1 above shows a summary of the differences between Tek Recon and AppBlaster Guns. AppBlaster provides a Single Player option to play, while Tek Recon does not. Nevertheless, Tek Recon provides Multi Player, Additional Game Modes, Physical Bullets, Online Scoring and Free Starting App, while AppBlaster does not provide Multi Player options, nor Physical Bullets. Thus, despite the impossibility of playing alone, Tek Recon provides wider options and possibilities to increase the game play experience.

The primary difference between the two products is observed by their basic gameplay functionality. Tek Recon is a primarily multiplayer experience whereas the AppBlaster is mostly a single player experience. Players with a Tek Recon gun are encouraged to invite their friends into teams and play together, while AppBlast-

Table 1 A simple comparison between the Tek Recon and AppBlaster guns

	Tek Recon	AppBlaster
Single player	No	Yes
Multi player	Yes	No
Additional game modes	Yes	Yes
Physical bullets	Yes	No
Online scoring	Yes	Yes
Free starting app	Yes	Yes
Devices	iOS/Android	iOS/Android

er's applications focus on simple games that can be played by the user at home alone.

The differing priority of the toy manufacturers is interesting as it shows a distinct difference between their marketing plans. The Tek Recon application does not offer much value to the consumer as a standalone program, but their toy gun does. On the other hand, the AppBlaster's gun does not have as much value by itself compared to the Tek Recon gun, but their games can be played without the gun and as such offer more value in that regard.

The mobile applications can also be mix and matched between different toy guns. Compatibility tests were conducted by having a smartphone loaded with *Alien Attack*, an AppBlaster application depicted in Fig. 12, mounted onto a Tek Recon gun. Despite the AppBlaster expecting a two-pronged tap to trigger a gun, if the smartphone was positioned properly it is still possible to play the game using the Tek Recon gun.

Marketing Strategies for Toys and Mobile Applications

Since the manufacturers value the pieces of their product differently, it is reflected in their pricing strategies. The Tek Recon application is available for download free of charge, whereas most of the AppBlaster's games cost a small fee. By offering a variety of applications available for purchase, Appfinity, AppBlaster's developer, can continue to make profits on their product after point of sale.

For toys like the Tek Recon and AppBlaster guns, the toy developers usually make ensure their applications only work with their respective guns. This is a challenge for them because, otherwise, consumers may purchase a competitor's application for their toy or use their application on a rival toy. AppBlaster does this by having the application react to the two pronged trigger on their gun, but it is not distinctly different enough that a user could not play their games on a competitor's gun.

On the other hand, mobile application developers can take advantage of this and develop generic applications that work on any toy interface. These developers focus their business model by supporting as many toy platforms as possible, similar to third party video game developers. As more and more toys are designed with mobile

Fig. 12 A screenshot of *Alien Attack*, a game for the AppBlaster. (Google Play[b] 2014)

applications in mind, it will be interesting to see if any submarkets appear as a result of this growing trend. One of the benefits that digital applications offer over physical objects is that they have a very minimal cost of distribution of subsequent copies. Traditional add-ons to toys must be manufactured and then sent to a distribution facility which incurs expenses for the toy manufacturer.

Digital products have the vast majority of their production costs in the development of the software and do not cost much to distribution online. Once the application has been developed it can be reproduced and sold over the Internet without incurring a lot of expenses. Modifications to the product can also be done in a similar way. When using modern digital games and mobile applications, it is not uncommon for users to download and install new content or patches to expand or improve the software. Developers can monitor their product's usage and metrics over the Internet after point of sale. This data can be used to determine how their product is being used and whether or not any improvements can be made to the software to improve the play experience of the product. Updates can then be pushed out to improve the quality of their product, which is logistically impossible with a non-digital product. This digital distribution system also allows developers to release downloadable content (DLC), to further monetize their products.

Besides development characteristics of these applications, some privacy and security concerns deserve attention from organization stakeholders and from literature considering that all of these applications run on the user's own devices. As long as organizations are allowing access to their data and network from employees' mobile devices, a strategy named BYOD (Bring Your Own Device), privacy and security deficiencies on mobile applications can cause inappropriate access to valuable information. Whenever an employee has one of these applications installed on his mobile device, the enterprise will be under the risk of having its information stolen or inappropriately accessed.

Armando et al. (2014) show the *secure meta-market* (SMM) framework that protect BYOD applications against other application's security and privacy failures. The SMM framework ensures that registered devices only run compliant policies. SMM also disable and replaces other clients, since they are not considered a secure installation. In addition, the authors created the BYODroid, a SMM prototype that supports Android OS and BYOD security policies. However, besides being developed just for one mobile operating system, the BYODriod does not support context-dependent policies, nor identify when a new app is not in accordance to the defined policies.

Integration of Analog Components to Digital Toys

Video games have also been incorporating more physical interactions into their game experience. Since the 1980's, the creation of analog components have been a field of interest in the digital toys industry. At that time, the famous Atari XEGS console became popular with sales of an analog gun, known as XE Light Gun. According to AtariAge (2014), just two games took advantage of the analog gun, Sentinel and Shooting Arcade. The following figures show, respectively, the XE Light Gun model (Fig. 13) and an example screen from the Sentinel game (Fig. 14).

Among more recent developments, analog components that imitate musical instruments have been successful in sales. Many of these, such as the instrument peripherals in the *Rock Band* series, have a toy-like device which acts as an interface to the game. The toy guitar, shown in Fig. 15, and the toy keyboard, shown in Fig. 16, are examples of many toy instruments that players can play with.

Beside musical instruments controllers, racing wheel devices are also widely popular among analog components for digital toys. Figure 17 shows a racing wheel for the third generation of Sony PlayStation® video games. This component allows players to play both car games and motorcycles games by adapting the position of the peripheral. Racing wheels provide intuitive ways for enjoying racing games. Each possible position can also be seen in Fig. 17.

Following the success of analog components to digital toys, a large number of research has been done in this field to explore the benefits of playing in augmented

Fig. 13 Atari XE light gun controller. (AtariAge 2014)

Fig. 14 Example screen of Sentinel game. (AtariAge 2014)

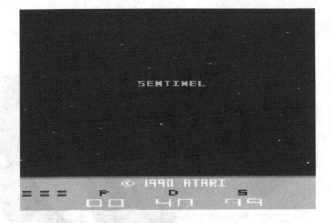

Fig. 15 The guitar controller from *Rock Band*. (Harmonix 2011)

Fig. 16 Rock Band keyboard for Xbox one. (Xbox 2014)

toy environments. For example, the research conducted by Hinske and Langheirich has shown that adding virtual elements to traditional toys brings forth the best of both worlds (Hinske and Langheinrich 2009).

The growing popularity of crowdsourcing platforms such as Kickstarter allows the proliferation of new peripherals to appear. The Delta Six controller for example, is an analog gun peripheral that can be used as an interface to popular first person shooter (FPS) games which delivers an experience akin to the Tek Recon blaster or AppBlaster. It is very likely that physical interaction through toy-like peripherals will become a large part of digital gaming within the next decade.

Fig. 17 PlayStation®Move racing wheel, retrieved from PlayStation®3 accessories website. (Playstation 2014)

Section 3. Security and Privacy Concerns

Location History and Data Tracking

It is not uncommon for mobile devices such as smartphones to be constantly connected to the Internet through a mobile data plan or a wireless Internet connection. According to Statista (2012), 96 % of all smartphone owners in the United States had a data plan in 2012. Having an Internet connection exposes the device to potential attacks and intrusions through the network which can result in many problems for the user (Koopman 2004). Different forms of attacks can include methods such as denial of service attack: attempting to make a machine or network resource unavailable, man in the middle: active eavesdropping on relays of data, and even spoofing: a person or program masquerades as another. Smartphones often log information about the user such as their movement patterns and chat logs which can be intercepted by malicious users or the government. This is a major security concern for toys that integrate mobile applications.

One of the most common types of logs a smartphone keeps track of is location history. Android phones already have the option to have their location history tracked by Google (Google 2014). If users opt to have the feature enabled, their GPS location will be periodically logged by their phone and sent over the network to Google. This is a feature that can be very useful as users can use the map in Fig. 18 to view their location history and keep of track of where they have travelled. This functionality is extended to find stolen or lost devices. However, if this data

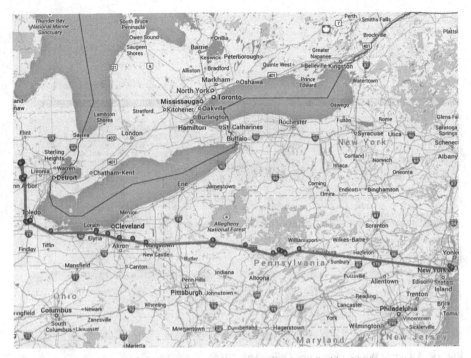

Fig. 18 Example of Google location history for a user. (Business Insider 2014)

is being sent over the network, it is also possible for it to be intercepted, which is a security concern. If a child were to be given an infected smartphone to play with, it is possible for hackers to activate the phone's GPS system to figure out where the child is.

For toy developers to properly integrate mobile applications with their appropriate product counterparts, it is imperative for them to ensure that their mobile device, application, and products are secure and do not put the consumer's information and privacy at risk. Privacy concerns exist whenever personally identifiable information or other sensitive information is collected or stored. This will be an ongoing problem because security measures will need to be consistently updated as malicious users obtain better tools and knowhow as to how to attack these devices and applications.

Although it is not limited to mobile gaming applications, there are several ways to lose important and confidential data. For example, some applications may contain hidden spyware, malware, phishing screens, and even background processes that track the user's data maliciously. Additionally, it is possible for the user's network traffic to be intercepted if their Internet connection is not adequately secure. For example, there are programs such as Wireshark which allows users to monitor their network traffic (Wireshark 2014). Unsecure data can be very easily be analyzed by using programs like Wireshark as it lays out the network data as shown in Fig. 19.

A study conducted by Websense (2012) shows that 59 % percent of their respondents report that employees circumvent or disengage security features, such as

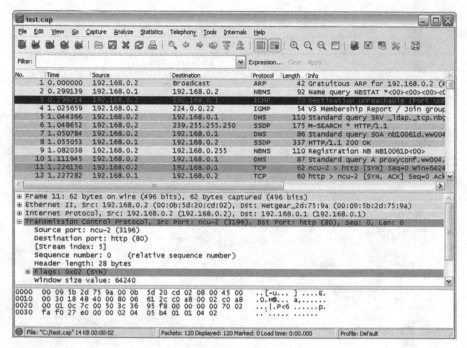

Fig. 19 A sample Wireshark capture. (Wireshark 2014)

passwords and key locks, on corporate and personal mobile devices. These organizations experienced an increase in malware (e.g. viruses, worms, spy-ware) infection due to insecure mobile devices. A recent study by Symantec (2012) showed that 15 % of social network users reported that someone had hacked into their profile. Security and privacy concerns are a barrier to entry for new toy developers to enter this industry because traditional toys do not require this type of security. As a result, additional costs must be incurred to develop and maintain the security framework.

Among this context, toy safety policies do not include security and privacy aspects yet. The most influent document about managing toy safety around the world, created by the International Organization for Standardization (ISO 8124-1 2012), reports policies and rules for:mechanical and physical properties; flammability;migration of certain elements;and swing, slides and similar activity toys. New parts are under preparation for future versions of this document, however, none of these new additions deal with security and privacy issues in toys.

Encryption and Data Security

For further information on mobile applications security and privacy online, potential communication and cryptographic flaws and mechanisms have to be con-

Fig. 20 AirAuth tool prototype. (Kratz and Aumi 2014)

sidered. Merriam Webster (2014b) defines encryption as the process "to change (information) from one form to another especially to hide its meaning." Encryption is an example of an effective way to ensure secure data transfer because if it is intercepted, the data is not easily decrypted to reveal its true content.

To ensure security on a mobile application, it must pass through both a handshake phase and a communication phase. The handshake phase's main purpose is to set up a connection between the server and the client, affirming each other's identity and confirming the symmetry secret key being used as a mechanism in communication phase. The communication phase transfers encrypted data between the server and clients using secret keys generated by the handshake phase. This phase is necessary to make identity authentication and provide secure communication, as it may contain confidential information. If lacking in security, potential unintended users can see private information, and therefore there exists the risk of a breach of privacy (Koopman 2004).

For the mobile application that is running on the user's smartphone, once authenticated, data is typically sent through a two way street, between both the application and the server. This data must be properly encrypted before being sent through the Internet to the server. Once the server receives the data, it must be decrypted for the data to be correctly read. If this data is not securely encrypted with a strong algorithm, it could be reverse engineered and therefore risking confidentiality and integrity. In addition, if not correctly secured, an attacker may not only be able to read information not intended for him/herself through transmission, but may even maliciously alter the data which may also risk the gathering of data stored on the mobile device if the application is not completely mutually exclusive with all other data on the mobile device itself.

As society becomes even more dependent on technology, memory retention can become an issue with mobile devices being used as an aid. Information such as usernames, passwords, and phone numbers are being stored on these devices. Without taking the proper security measures, developers risk compromising the security and privacy of their customer's confidential information. As a reaction against this gap, the interest in researches about gesture authentication has increase during the past few years. As an example, Krutz and Aumi (2014) are developing the AirAuth tool shown in Fig. 20 that aims to allow a user to authenticate using both biometrical and gesture analysis, based on an in-air gesture input.

Section 4. Discussion

Future Developments

As toy developers begin integrating additional digital elements into their toys, there will be many increasing digitally augmented toys in the future. With the rate at which digital technology is developing, the potential that exists in toys with mobile application integration will grow and allow developers to create content for their players that cannot simply exist today. In the near future, the distinction between a toy and a video game is likely to decrease as consumers demand more immersive play experiences.

Smartphones are getting increasingly powerful which will inevitably allow developers to leverage this power to create more ambitious and technically advanced pieces of mobile application. For example, imagine an augmented reality game that is constantly connected to a digital network system. People might be able to take a gun such as the Tek Recon blaster and mount their smartphone onto it similar to what they can do today.

The front end system of the mobile application merely acts like the HUD and displays information to the player about the current game state. However, the smartphone will also do a lot of data logging and tracking in the background. By using the camera built into the smartphone, the application will be able to constantly scan and process the player's environment. The application can also use other components to track and log the player's position and orientation. All of the data can be collected and then sent to an online server.

With the data gathered by all of the players in the match, a 3D reconstruction of the real life game environment can be simulated by the game system. This reconstruction can be generated through a variety of techniques. For example, each player's camera can be used to obtain data on the physical layout of the environment the match is played on. This data can be used to stitch together an approximate representation of the playing field.

The constant stream of data from the player's devices allows a replay of the match to be saved online. Currently, popular multiplayer games such as *StarCraft II*, support a similar replay system where the stored game data can be viewed in an interactive video form that can be used to analyze the player's strategies (Blizzard Entertainment 2013). With a system like this in place, each player will be able to log onto their gaming consoles or computers and review their game tactics and strategies by analyzing this replay or raw data like in Fig. 21. Players will also be able to share this replay with their friends or show it off online.

Each player will be represented in the virtual replay by their customizable 3D avatar and thus a community can be built around this hypothetical gaming system. By completing achievements, the player can earn new levels of customization that makes their avatar stand out. Since this avatar represents the player in the 3D reconstructions of each game they play, having a character equipped with rare customizations can be a symbol of skill and high social status. Essentially, the toy and

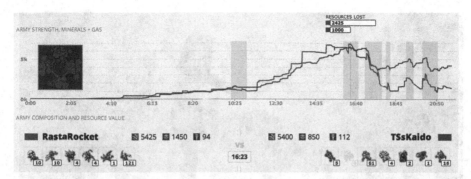

Fig. 21 A visualization of data extracted from a *StarCraft II* replay. (Blizzard Entertainment 2013)

smartphone acts as an immersion interface into a first person video game where the player can fully experience what it is like to play in a video game. This type of game is also unique because it gives the player full freedom in the way they want to play the game, unlike in traditional shooter games where the player is limited to the maps and mechanics that are given to them by the developers.

Research has already been done in this area to prove that it is possible to generate 3D environments from the real world. For example, images can be analyzed to determine "where points in one image move to in another image" (Rander et al. 1997) which can be used to orient and place objects in the scene. Depth information and GPS coordinates from the player's smartphone can also be utilized to better construct the digital environment. However, this technology is not yet readily available, or sufficient, on a consumer level. Once technology has advanced to the point where this is possible, it unlocks many more opportunities for unique virtual experiences in the physical world. Toys can be augmented in such a way to deliver unparalleled play experiences.

Other ways that this technology can be leveraged would be to use existing VR technology such as Computer Assisted Virtual Environments (CAVE) like the one in Fig. 22. CAVEs provide the illusion of reality to the users inside them as they project high resolution images which are brought to life with 3D glasses and head tracking (EON Reality 2014). This type of environment provides the user with a complete visual immersion, but it still lacks haptic and physical feedback. However, toys that are equipped with digital elements can be integrated with the system to provide that extra layer of immersion to improve their VR experience. This can also be done with other types of VR technologies such as head-mounted displays such as the Oculus Rift.

Toy Guns in Brazil

In Brazil, the toy industry has been growing significantly during the last decade. According to ABRINQ (2013), the Brazilian toy association, the toy industry has

Fig. 22 The EON cube demonstrating CAVE technology. (EON Reality 2014)

earned more than R$4B in 2013 and has provided 27,000 jobs in 2012. However, the toys industry in Brazil is unfortunately facing some problems with public safety.

Since 2003, the Brazilian Federal Government prohibits the production and sale of replica toy guns, toys that are replicates of real guns (see an example in Fig. 23). This prohibition is guaranteed by Federal Law number 10,826, from December 22, 2003, and the justification for this is that this kind of toy was used to practice crimes (Palácio do Planalto Presidência da República 2014). However, this action was not sufficient to avoid and stop crimes made using toys.

According to the *Instituto Sou da Paz* (2013), between 2011 and 2012, 37.6 % of guns used in crimes inside the state of São Paulo were toys, simulacrum or pressure guns. In 2014, the state of São Paulo approved a project of law to prohibit production and sale of toy guns (even those that are not replicates of real guns, see Fig. 24), law project number 15,301/2014 (*Assembléia Legislativa do Estado de São Paulo* 2014) is aiming to avoid these kind of crimes and argues that the use of toy guns also increases the aggressive behavior in children.

In 2013, the discussion about toy guns enhancing aggressive behavior in kids also led the Brazilian Federal District to enforce a law to prohibit sales and produc-

Fig. 23 An example of replica toy gun. (Daily Mail 2010)

Fig. 24 An example of toy
gun not considered a replica.
(Tek Recon[b] 2014)

tion of these kind of toy guns. The refereed law is Distrito Federal (Federal District) state Law number 5,180 (*Câmara Legislativa do Distrito Federal* 2013) from September 20th, 2013.

According to the facts described above, toy guns are becoming synonymous of crime in Brazil. This means that more laws and prohibitions of these kind of toys will probably be done, impacting related jobs in toy factories and stores and also diminishing the economic impact of the toy industry in Brazil. Nonetheless, technology infused toys are seen as a potential way to manage Brazil out of these complications. Thus, research about technology applications in this case must be done in order to find solutions that lead to safer usage of toy guns and better public safety in Brazil.

Section 5. Conclusion

The toy industry has existed for many millennia and changes have occurred as trends come and go. As technology continues to develop, other industries and fields must adapt to stay relevant and competitive.

By analyzing the cases found in this chapter, it is shown that the integration of smartphone applications and digital technology is a recent trend in the industry. Digital technology allows toy manufacturers to augment the overall play experience of toys and to deliver a product that is more dynamic. Digital technology permeates the lives of billions of people around the world; it is expected that toys are being designed with digital technology in mind.

As toy developers continue to integrate digital technology such as smartphones into the designs of their toys, new trends will arise that will shape the future of the toy industry. New and exciting ways for users to play with toys have been created which has diversified the products available for consumers to purchase.

Also, this chapter helped to understand reasons why it is important to attend for security and privacy concerns about mobile applications developments, so that the integration toy-technology must be done well. The view of future trends shows how 3D technologies will likely continue to be more popular in toys in coming years.

References

ABRINQ. (2013). A força do brinquedo.

Amazon. (2014). Amazon Best Sellers. Retrieved August 27, 2014, from Amazon.com: http://www.amazon.com/Best-Sellers-Cell-Phones-Accessories/zgbs/wireless/ref=zg_bs_unv_cps_1_2407747011_1

Apptoyz. (2014). appBlaster v2. Retrieved August 16, 2014, from Apptoyz: http://www.apptoyz.com/shop/appblasterv2/

Armando, A., Costa, G., Verderame, L., & Merlo, A. (2014). Securing the "Bring Your Own Device" Paradigm. *Computer, 47*(6), 48–56.

Assembléia Legislativa do Estado de São Paulo. (2014). Lei n° 15.301, de 12/01/2014. Retrieved September 24, 2014, from Assembléia Legislativa do Estado de São Paulo: http://www.al.sp.gov.br/norma/?tipo=Lei&numero=15301&ano=2014

AtariAge. (2014). Controllers—XE Light Gun. Retrieved September 16, 2014, from AtariAge: https://atariage.com/controller_page.html?SystemID=2600&ControllerID=13

Bandai. (2014a). SUPER BEST 変身ベルト DXオーズドライバー (Super Best Henshin Belt DX OOO Driver). Retrieved March 13, 2014, from Bandai: http://www.bandai.co.jp/catalog/item/4543112826787000.html

Bandai. (2014b). レジェンドライダーシリーズ　変身ベルト 仮面ライダー新1号 (Legendary Rider Series Henshin Belt Kamen Rider Ichigo). Retrieved March 13, 2014, from Bandai: http://www.bandai.co.jp/catalog/item/4543112550408000.html

Bandai Namco Games Inc. (2013, August 20). Tamagotchi L.i.f.e. Retrieved May 22, 2014, from iTunes: https://itunes.apple.com/ca/app/tamagotchi-l.i.f.e./id608818625?mt=8

Bellotti, F., Kapralos, B., Lee, K., Moreno-Ger, P., & Berta, R. (2013). Assessment in and of Serious Games: An Overview. Advances in Human-Computer Interaction.

Blizzard Entertainment. (2013, May 7). StarCraft II Patch 2.0.8 Replay File Enhancements. Retrieved August 16, 2014, from Battle.Net: http://us.battle.net/sc2/en/blog/9669862

Bradford, K. T. (2014, February 18). Techlicious 2014 Best of Toy Fair Awards. Retrieved August 16, 2014, from Techlicious: http://www.techlicious.com/blog/techlicious-2014-best-of-toy-fair-awards/

Brian, Matt. (2014, August 12). Toys that let you send voice messages to your kids are coming to the UK. Retrieved August 16, 2014, from Engadget: http://www.engadget.com/2014/08/12/toymail-voice-message-toy-kids-uk/?ncid=rss_truncated

Burton, A. (1997). Design History and the History of Toys: Defining a Discipline for the Bethnal Green Museum. Journal of Design History, 1–21.

Business Insider. (2014). Use This Trick To See A Map Of Everywhere Google Knows You've Been. Retrieved September 24, 2014, from Business Insider: http://www.businessinsider.com/see-what-google-knows-about-my-location-2014-8

Câmara Legislativa do Distrito Federal. (2013). Processo Legislativo—Norma Jurídica—Ficha Técnica. Retrieved September 24, 2014, from Câmara Legislativa do Distrito Federal: http://legislacao.cl.df.gov.br/Legislacao/consultaTextoLeiParaNormaJuridicaNJUR-10813!buscarTextoLeiParaNormaJuridicaNJUR.action

Daily Mail. (2010). Child's toy gun sparks panic at bank branch. Retrieved September 24, 2014, from Daily Mail: http://www.dailymail.co.uk/news/article-1292416/Toy-gun-dropped-child-sparks-panic-bank-branch.html

de Lima Salgado, A., & Freire, A. P. (2014). Heuristic Evaluation of Mobile Usability: A Mapping Study. In *Human-Computer Interaction. Applications and Services* (pp. 178–188). Springer International Publishing.

EON Reality (2014). EON Icube. Retrieved August 4, 2014, from EON Reality: http://www.eon-reality.com/eon-icube/

Fightglobal. (2009). Apollo @40. Retrieved February 27, 2014, from Flight Global: http://www.flightglobal.com/page/Apollo-40th-Anniversary/

Fundação Hospital Amaral Carvalho. (2014). Retrieved August 25, 2014, from Fundação Hospital Amaral Carvalho: http://www.amaralcarvalho.org.br/amaralcarvalho/pt/noticia/visualizar/codnoticia/781/no-hospital-amaral-carvalho-ursinhos-com-tecnologia-inovadora-levam-mensagens-de-familiares-e-amigos-as-criancas-com-cancer.html

Google. (2014). Location in Google settings. Retrieved August 2, 2014, from Google: https://support.google.com/accounts/answer/3118687?ref_topic=3100928&hl=en

Google Play[a]. (2014). Sphero. Retrieved September 24, 2014, from Google Play: https://play.google.com/store/apps/details?id=orbotix.sphero

Google Play[b]. (2014). apptoyz Alien Attack. Retrieved September 24, 2014, from Google Play: https://play.google.com/store/apps/details?id=com.AppToyz.AlienAttack

Harmonix. (2011, November 23). Rock Band instrument prototype & insider stories. Retrieved August 3, 2014, from Harmonix: http://www.harmonixmusic.com/blog/rock-band-instrument-prototypes-insider-stories/

Hinske, S., & Langheinrich, M. (2009). An Infrastructure for Interactive and Playful Learning in Augmented Toy Environments. Pervasive Computing and Communications, 1–6.

Huang, W.-C., Lu, T., & Fang, W.-C. (2011). An Innovative Interface Design with Smart Phone for Interactive Computer Game Applications. Games Innovation Conference (IGIC), 62–65.

Instituto Sou da Paz. (2013). DE ONDE VÊM AS ARMAS DO CRIME: ANÁLISE DO UNIVERSO DE ARMAS APREENDIDAS EM 2011 E 2012 EM SÃO PAULO.

International Data Corporation. (2013, August 7). Apple Cedes Market Share in Smartphone Operating System Market as Android Surges and Windows Phone Gains, According to IDC. Retrieved January 4, 2014, from IDC: http://www.idc.com/getdoc.jsp?containerId=prUS24257413

ISO 8124-1. (2012). Safety of toys. International Standard.

Koopman, P. (2004). Embedded System Security. Computer, 95–97.

Kratz, S., & Aumi, M. T. I. (2014, April). AirAuth: a biometric authentication system using in-air hand gestures. In CHI'14 Extended Abstracts on Human Factors in Computing Systems (pp. 499–502). ACM.

Maspero, G. C. (1895). Manual of Egyptian Archaeology and Guide to the Study of Antiquities in Egypt. Project Gutenberg.

Merriam-Webster. (2014a). Augmented Reality. Retrieved March 13, 2014, from Merriam-Webster: http://www.merriam-webster.com/dictionary/augmented%20reality

Merriam-Webster. (2014b). Encrypt. Retrieved August 16, 2014, from Merriam-Webster: http://www.merriam-webster.com/dictionary/encryption

Moore, G. E. (1965). Cramming More Components onto Integrated Circuits. Electronics, 82–85.

Nintendo. (2014). AR Cards. Retrieved May 22, 2014, from Nintendo 3DS Official Site: http://www.nintendo.com/3ds/ar-cards

Nokia Developer. (2014). File:Unity-AfterNewEnvironmentCreated.png. Retrieved September 24, 2014, from Nokia Developer: http://developer.nokia.com/community/wiki/File:Unity-AfterNewEnvironmentCreated.png

Palácio do Planalto Presidência da República. (2014). LEI No 10.826, DE 22 DE DEZEMBRO DE 2003. Retrieved September 24, 2014, from Palácio do Planalto Presidência da República: http://www.planalto.gov.br/ccivil_03/leis/2003/l10.826.htm

Rander, P., Narayanan, P. J., & Kanade, T. (1997). Virtualized Reality: Constructing Time-Varying Virtual Worlds from Real World Events. VIS '97 Proceedings of the 8th conference on Visualization '97, 277–ff.

Nayak, M. (2013, June 10). FACTBOX—A look at the $66 billion video-games industry. Retrieved August 2, 2014, from Reuters: http://in.reuters.com/article/2013/06/10/gameshow-e-idINDEE9590DW20130610

Playstation. (2014). PlayStation®Move Racing Wheel. Retrieved September 16, 2014, from Playstation: http://us.playstation.com/ps3/accessories/playstation-move-racing-wheel-ps3.html

Samsung. (2013). Samsung Galaxy S4 Specifications. Retrieved February 27, 2014, from Samsung: http://www.samsung.com/global/microsite/galaxys4/

Statista. (2012). Share of Smartphone Owners with a Data Plan in 2012, by Country. Retrieved from Statista: http://www.statista.com/statistics/262757/share-of-smartphone-owners-with-a-data-plan-by-country/

Stonehouse, A. (2014, February 27). User interface design in video games. Retrieved August 3, 2014, from Gamasutra: http://gamasutra.com/blogs/AnthonyStonehouse/20140227/211823/User_interface_design_in_video_games.php?print=1

Tek Recon. (2014). Tek Recon. Retrieved May 22, 2014, from Youtube: https://www.youtube.com/watch?v=oRRIu-eSVkA

Tek Recon[b]. (2014). Tek Recon. Retrieved September 24, 2014, from Tek Recon: www.tekrecon.com

Symantec. (2012, September 5). 2012 Norton Study: Consumer Cybercrime Estimated at $110 Billion Annually. Retrieved August 6, 2014 by Symantec: http://www.symantec.com/about/news/release/article.jsp?prid=20120905_02

Toy Industry Association. (2013, February 11). Toy Trends for 2013 Announced by Toy Industry Association. Retrieved December 3, 2013, from Toy Fair: http://www2.toyassociation.org/AM/Template.cfm?Section=TF_News&CONTENTID=18594&TEMPLATE=/CM/ContentDisplay.cfm

Websense, Inc. (2012, February 29). Survey Says 51 Percent of Organizations Lose Data through Mobile Devices, 59 Percent of Employees Dodge Security Controls. Retrieved August 6, 2014, from PRNewswire: http://www.prnewswire.com/news-releases/survey-says-51-percent-of-organizations-lose-data-through-mobile-devices-59-percent-of-employees-dodge-security-controls-140850593.html

Wireshark. (2014). About Wireshark. Retrieved August 16, 2014, from Wireshark: https://www.wireshark.org/about.html

Xbox. (2014). Rock Band Keyboard. Retrieved September 16, 2014, from Xbox: http://marketplace.xbox.com/en-us/Product/Rock-Band-Keyboard/00001000-a683-b283-ca68-b52045410914

Emerging Human-Toy Interaction Techniques with Augmented and Mixed Reality

Jeff K. T. Tang and Jordan Tewell

Abstract In this book chapter, we will review the emerging technologies that promote extensive interactions between toys and their players. Cutting-edge display technologies have taken a major role in human-toy interaction. In particular, reality-virtuality technologies such as Augmented Reality (AR) and Mixed Reality (MR) have been adopted in digital entertainment and physical toys. With advances of portable and wearable devices, reality-virtuality technologies have become more popular and immersed into our daily life. Various interaction techniques are identified such as depth sensors and haptic devices. We will introduce example technologies, devices and products of the above technologies. However, displayed/projected virtual objects cannot give the user a sense of touch. Additional apparatus such as haptic styli are necessary to interact with the virtual objects. We foresee that in the near future, more kinds of virtual senses (such as taste) could be simulated and become part of the toy.

Keywords Augmented Reality · Interaction Techniques · Feedback · Virtual Senses · Toy Computing

Introduction

Traditional toys were touchable physical objects and most of them were silent, i.e. unable to respond to human instructions. In order to make them more fun, players have to manually manipulate the toys (such as dolls) in order to simulate their responses. They required additional imagination and storytelling skills. Hence, they were not interactive at all.

With the advances of computing technologies in the twenty-first century, toys have jumped out of the box and became more interactive for players. Many modern

J. K. T. Tang (✉)
School of Computing and Information Sciences, Caritas Institute of Higher Education,
Hong Kong
e-mail: jtang@cihe.edu.hk

J. Tewell
School of Mathematics, Computer Science & Engineering, City University London,
United Kingdom
e-mail: jordan.tewell.1@city.ac.uk

© Springer International Publishing Switzerland 2015   77
P. C. K. Hung (ed.), *Mobile Services for Toy Computing,* International Series on
Computer Entertainment and Media Technology, DOI 10.1007/978-3-319-21323-1_5

toys are computerized or programmable and hence they are able to interact and communicate with human beings. Amongst the emerging technologies, Augmented Reality (AR) is becoming more popular in toys and games. It provides the capability for toys to fuse the matters of the virtual world into the real world.

Toy Computing is a novel study of computing technologies that involve the manufacturing and the game play design of toys. Novel technologies could give new experience to toy players and hence evolve into new interaction styles. Hence, it is necessary to carefully design the interface where the human and the toy interact, no matter their interactions occurring in the physical (real) and the virtual domains. We introduce a novel term "Human-Toy Interaction" to describe this specialized study.

AR Toys provide a rich Human-Toy Interaction experience. Novel display technologies such as head mounted displays (HMD) are able to overlay additional information onto a real scene/environment in holographic form, and hence making the virtual objects seamlessly immersed in the real world. A smartphone monitor is the most popular AR display device. AR is quite different from Virtual Reality (VR) as the latter is totally virtual and immersive. Nowadays, HMD technology has become more light-weight and wearable such as the Google GLASS (2014) and Microsoft HoloLens (http://www.microsoft.com/microsoft-hololens/en-us) so people can use it anytime and anywhere. However, a more natural user interface is demanded because such devices are not totally "hands-free". Moreover, virtual objects cannot give the player a sense of touch. It is necessary to simulate haptic and virtual senses in order to make the interaction more realistic. In this chapter, we are going to reveal state-of-art technologies that possibly tackle the above needs.

This chapter is organized as follows. In Section 2, we will introduce various toys and games ranging from real to virtual domains, followed by the detailed explanation of augmented reality (AR) techniques. Section 3 will describe the recent advances of display technologies. Section 4 will describe how humans interact with virtual toys in a mixed reality environment, followed by the description of the state-of-art haptic and virtual senses techniques.

Toys and Games on the Reality-Virtuality Continuum

In this section, we will introduce examples of toys in different technologies on the *Reality-Virtuality Continuum* of advanced computer interfaces by Milgram and Kishino (1994) as shown in Fig. 1. It describes a range of reality-virtuality technologies between the completely virtual and the completely physical.

The reality-virtuality technologies are briefly introduced as follows: Virtual Reality (VR) requires users to be immersed into a virtual environment and interact with virtual toys in a purely virtual domain (Van Dam 2000). Augmented Virtuality (AV) describes the adoption of real world information (such as movements, orientations) into the virtual world (Regenbrecht et al. 2003). Augmented Reality (AR) describes the capability of overlying additional graphical and textual information onto the toy (or its surroundings) via a display device (Azuma et al. 2001).

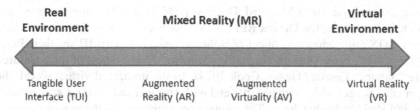

Fig. 1 Our overall approach

Mixed Reality (MR) refers to the hybrid of AR and AV technologies, where physical objects (or humans) and virtual objects (or characters) exist in the same place and their real-time interactions can occur in both real and virtual worlds (MR). Tangible User Interface (TUI) describes the interface where the player interacts with digital content via manipulating physical objects. It is the technique that provides the user with a sense of touch or feedback in addition to Augmented Reality (Ishii 2008).

These reality-virtuality technologies not only make toys more fun, but also motivate players to learn new knowledge or skills. In the following subsections, we will introduce the toys and games of each reality-virtuality technology.

Virtual Reality (VR)

VR can be either immersive or semi-immersive (Van Dam 2000). Immersive VR always uses a special type of display device that completely covers the player's field-of-view (FOV). For example, the Cave Automatic Virtual Environment (CAVE) is a room-sized cube of projectors where the walls are screens (Cruz-Neira 1993). However, CAVE is very expensive and it needs to be installed in a large area. Hence, it is unsuitable for domestic toys. However, it is large enough for large-scale installations such as a virtual museum as shown in Fig. 2 (left).

Semi-immersive VR covers only a part of a player's FOV. It does not require a large area for image projection and hence more suitable for domestic toys and

Fig. 2 An immersive virtual reality museum with CAVE Graphics (Kolb 2008) (*left*) and a semi-immersive video game with Oculus Rift (Game News 2013) (*right*)

games. Nowadays, Head-Mounted Display (HMD) has become slimmer and lighter. For example, the Oculus Rift (2014) has been supported by many video games. SONY introduced "Project Morpheus", which is a HMD for their Play-Station game console (Kelion 2015). Fig. 3 shows the snapshots of their featured game "Summer Lesson" (James Cook 2014). In the immersed virtual world, the player can interact with a virtual girlfriend with simple head gestures such as nod-ding and shaking his/her head. The interaction is natural as the player does not require control devices.

Augmented Virtuality (AV)

As an intermediate case in the Virtuality Continuum, Augmented Virtuality (AV) refers to the dynamic integration and interaction of virtual spaces with physical objects or people in real-time. For example, video/data points can be streamed from the real world through a video camera that drives characters/objects in the virtual space. An example of this is motion capture technology.

The Optical Motion Capture system needs to be setup in a large area. It is very accurate, though expensive, so it is good for detailed analysis of human motion in a laboratory setting. Fig. 4 (left) demonstrates a serious game that guides a player to dance properly with the aid of optical motion capture (Chan et al. 2011).

The Microsoft Kinect is capable of estimating the skeleton of a player's body (Kinect for Windows 2015). It is regarded as a cheaper motion capture solution (about US$ 100) and is affordable for most domestic users, in particular, video game players. Fig. 4 (right) shows two people playing an adversarial game in which their avatars are controlled by their full body motions captured by Kinect.

Fig. 3 The VR game "Summer Lesson" developed by SONY (James Cook 2014). The player immersed in virtual world with the SONY morpheus display (*left*) and she can interact with the virtual character by simple head gestures (*right*)

Fig. 4 A serious game for learning dance with optical motion capture system (Chan et al. 2011) (*left*) and an Xbox sport game with Kinect depth sensors (Kinect Games 2015) (*right*)

Augmented Reality (AR)

Augmented Reality (AR) refers to the superimposition of information onto a real scene. It enhances the functionality of toys and makes them more interesting as well as educational. The player could view the real world environment supplemented by computerized sensory input such as audio, graphics, video, GPS, etc. A display (such as a HMD) is needed to view such kind of supplementary information (Billinghurst et al. 2005).

The game "appBlaster" is a plastic toy gun with AR capability (AppToyz Blaster 2014). It allows the player to put a smartphone on it in order to capture the real scene and display a "killing field" as shown in Fig. 5 (left). The player needs to install the game app provided, which randomly displays the aliens onto the real scene captured by the phone's camera. The player needs to collect the weapons and

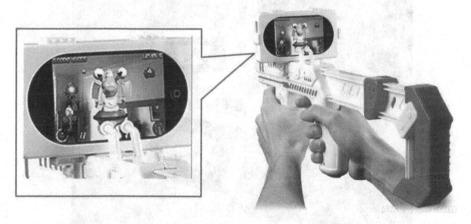

Fig. 5 The "appBlaster" AR toy gun. (AppToyz Blaster 2014)

powers displayed around the environment, track down the aliens, and shoot them. The "fire" button can trigger different touch patterns to the smartphone screen. The player can perform different game functions by changing the gesture (i.e. the multitude of the gyroscope).

It would be more useful if the augmented information is shown in a specific place accurately rather than just random. Hence, advanced AR techniques are being developed to capture the real world positions and orientations of desirable places. These can be classified by marker-based methods (using paper cards with special patterns as an agent) and markerless methods (not using an AR marker as agent but detecting the coordinates and properties of real objects in real-time).

Marker-based AR

Typical capture and recognize AR markers are shown in Fig. 6. AR markers are printed cards that are always a square pattern with a bolded outline. Traditionally, AR markers are in black-and-white. Currently, there are many AR software development kits (SDKs) available for researchers/developers. Different SDKs have their own characteristics, and the choice of the SDK depends on the developers' needs. Some advanced SDKs are able to recognize colors as well as complicated textures. Colored AR markers are always used as the product packaging hence it is very good for advertising purposes.

The ARToolKit developed by Kato (AR Toolkit 2002; Kato and Billinghurst 1999) is the earliest SDK for building AR applications. It is capable of tracking the position and orientation of printed markers, and can place virtual images or 3D objects on top of the markers. In particular, 3D models can be rendered by importing graphics libraries such as OPENGL ES for mobile devices. ARToolKit supports AR

Fig. 6 Typical AR markers (getting started—ARToolworks support library 2014)

software development on both desktop and mobile platforms. NyARToolkit (http://nyatla.jp/nyartoolkit/wp/) is derived from ARToolKit that provides object oriented API libraries for AR software development. It adopts fast fitting and fast labeling methods that enhances the AR marker recognition. It targets mobile application development and supports a wide range of programming languages such as Java, C#, and Processing. Both ARToolKit and NyARToolkit are freeware and publicly available for educational and research use.

Qualcomm's Vuforia augmented reality SDK (http://www.qualcomm.com/Vuforia) is a commercial tool for developing AR software products. It is more powerful than ARToolKit and NyARToolkit. It supports C++, Java, and Android. Also, it can be integrated with *Unity*, which is a 2D/3D game development platform more robust than OpenGL ES. Hence, it renders 3D models and animations more professionally, and the developer would have less struggle in building AR software with 3D modeling. In addition, it can track various types of AR markers from simple black square patterns to more complicated colored pictures. This SDK also provides a web tool for developers to create and evaluate their own AR markers to be used in their applications.

In recent years, AR toys have become more popular. The AR figure *AR* is developed by Geisha Tokyo Entertainment (2009) is an interactive virtual girl. Figure 7 shows the virtual girl AR is sitting (i.e. rendered) on a physical cube of AR markers. The player can interact with the virtual girl through different rods or cards printed with AR markers. For example, the player can give her a teddy bear using a "gift" marker.

Figure 8 shows a concept design by the French designer Lasorne (2009). He imagined that future game play of augmented reality toys will become tangible, i.e. offers the players both virtual and physical experiences. In the figure, the player wore a glasses-like Head-Mounted Display, which rendered additional information (such as the health point of a LEGO figure) and augmented functions in terms of graphics on an AR marker. Virtual weapons will be loaded by manipulating the augmented functions. By moving the LEGO figure, the target point of the weapon could be adjusted.

Fig. 7 The virtual figure AR is rendered on an AR marker (Geisha Tokyo Entertainment, Inc. 2009) (*left*). The player can interact with AR is with different patterned markers

Fig. 8 A player plays with a toy figure while its health point (HP) is shown (*left*). A virtual gun and its targeting point are visualized on the screen (*right*)

Markerless AR

The markerless AR method aims at recognizing static (or even moving) objects of a scene in real-time, which is more powerful than the marker-based AR method that only recognizes known marker patterns. The Scale-Invariant Feature Transform (SIFT) proposed by Lowe (2004) is a useful method to recognize the objects appearing in a colored scene.The features extracted by SIFT are invariant to scaling and orientation, so it is capable to recognize objects in the scene even though they are partly occluded or cluttered with each other. The AR function can be enabled by placing the virtual images/videos on the flat surfaces of the recognized objects.

The operation of SIFT is based on searching key points within the image of a scene. In order to extract the key points, a training step is required in which the Euclidean distances of the new incoming scene and each of the pre-stored images are evaluated. As a result, the best matches in the new image are identified as key points. Some researchers identify the key points by determining the Difference of Gaussians (DoG) i.e. the difference of several Gaussian blurred images.

Eddington demonstrated how to extract key points in a scene and render a virtual clock on a poster recognized by SIFT as shown in Fig. 9 (right) (Matas et al. 2002). In addition, he enhanced the accuracy of SIFT by adopting the Maximally Stable Extremal Regions (MSER) method (http://users.csc.calpoly.edu/~zwood/teaching/csc572/final11/geddingt/), which could find the corresponding regions of two images of different scales and orientations.

Modern educational toys enable players to learn new knowledge or attain new skills in the AR environment. For example, AR Flashcards (Mitchlehan Media, LLC 2013) as shown in Fig. 10, provides an intuitive and interactive way of learning. The player can view the 3D model of an animal by poking at the printed flashcard on the device's screen. He or she can tap on the animal's body to hear the pronunciation of the animal's name and hear how it roars. The beautiful graphics and sound effects made learning more attractive and fun for small children. Figure 11 shows another similar app called the AR Dinosaur (2014). It displays different types of live animated dinosaurs on different AR marker cards.

Fig. 9 An example AR application by SIFT algorithm (Matas et al. 2002). The key points located by SIFT (*left*). The virtual object displayed on the identified real object (*right*)

Fig. 10 AR flashcards (Mitchlehan Media, LLC 2013)

Fig. 11 AR dinosaur (2014)

Fig. 12 The badger sends message (*left*) and the deer retweeting badger (*right*)

Suwappu is a series of markerless AR toys prototyped by London designers (ONeal 2011). It consists of a mobile app and some animal character figurines. The player needs to place an animal character figurine on a plain surface such as a table. The Suwappu app can recognize the animal and displays a holographic world for the animal on the screen as shown in Fig. 12. The player can manipulate the character and messaging with other characters. This function makes it a potentially useful toy for digital storytelling.

Mixed Reality (MR) and Tangible User Interfaces (TUI)

Mixed reality (MR) refers to the *hybrid reality* that integrates the physical and virtual objects and hence a new environment and visualization are created. It is a mixture of augmented reality and augmented virtuality technologies so MR enables the player to interact with both physical and virtual objects in real-time.

Sphero is an app-controllable robotic ball that is able to roll and change colors. The outlook and inner composition is shown in Fig. 13. Figure 14 shows the Mixed-Reality Pong with Sphero (2015). The players need to control the physical Sphero ball rolling through the opponent. The opponent has to move him/herself in order to make the virtual paddles on their screens bounce back the ball[1].

Tangible User Interface (TUI) provides another MR interaction experience, which uses physical objects to both represent and interact with computer generated information (Ullmer and Ishii 2001). Ichida et al. designed the ActiveCube,in which the user can use the physical cubes to construct a virtual model in real-time as shown in Fig. 15 (left) (Ichida et al. 2004). The QUMARION by SoftEther (2011) is a tangible humanoid input device. Through moving the limbs of the humanoid, CG artists and animators could produce different postures for a 3D model in a quicker and more natural way. Figure 16 shows how the CG posture (right) is rendered in correspondence to the posture of the physical humanoid input.

---

[1] Go Sphero. (29th Sep. 2011). *Mixed-Reality with Sphero* [Video file]. Retrieved from. https://www.youtube.com/watch?v=rm00f8jlXpg

Fig. 13 The Sphero robotic ball and its inner composition

Figure 17 shows an interactive toy called Tiggly Shapes (Kidtellect 2012). It consists of an iPad education app with four physical geometric shapes: square, circle, triangle and star. The player is instructed to place the correct shape on the iPad screen surface, in order to unlock a corresponding virtual object in the game. There are currently three Tiggly apps available: one for drawing, one for shape matching, and one that brings in animated characters for the browser. Manipulating physical shapes can help children to develop fine motor skills and spur their creativity and spatial reasoning.

The AppMATes is a game developed by Disney (Disney/Pixar 2012). It consists of an iPad game app and toy cars themed around the movie "Cars 2" by Pixar. As shown in Fig. 18, the user holds a toy car in the center of the iPad screen surface, and the scene will change that to simulate the car driving in the virtual world.

In the past, people looked into virtuality more as a technology to simulate the real world in virtual domain. Hence, immersive systems were introduced. Nowadays, people shift back to the real domain, such that the virtual objects act like real things and are superimposed onto the real environment. In mixed reality, people can interact with the virtual world with real objects, which is more direct and intuitive.

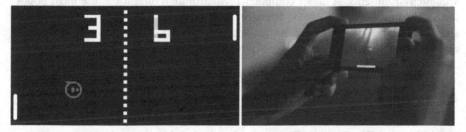

Fig. 14 The traditional pong video game interface (*left*) and the mixed reality pong interface (*right*). The light dot is the physical sphero robotic ball captured by the camera (http://www.gosphero.com/blog/mixed-reality-pong-with-sphero/)

Fig. 15 ActiveCube. (Ichida et al. 2004)

Fig. 16 QUMARION. (Soft-Ether 2011)

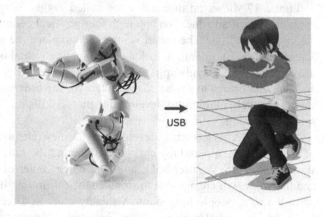

Display Technologies

Novel display technologies are vital for displaying additional information in real scenes/environments. A tablet computer or smartphone monitor is the most popular AR display device. With the advances of display devices, the virtuality can be shown in the real world easily. In this section, we will discuss how these devices allow the development of a new type of toys.

Head-Mounted Display (HMD)

Head-Mounted Display (HMD) is a display device that the user can wear on the head. Sutherland is the pioneer of HMD system and he created *The Sword of Da-*

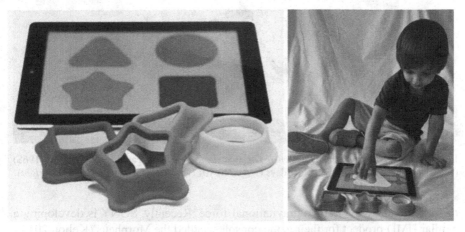

Fig. 17 Tiggly stamps produced by tiggly shapes. (Kidtellect 2012)

Fig. 18 The Disney's
AppMATes(TM) toy car.
(Disney/Pixar 2012)

mocles[2] in 1968, which is a helmet-like binocular HMD that consists of two little display optics in front of the eyes as shown in Fig. 19 (left) (Sutherland 1968). It is capable of producing a 3D vision as the two monitors could display different vision signals. This system can display simple wireframe geometry shapes and overlay them on the real scene recorded by the camera. Hence, it is regarded as the world's first virtual reality and augmented reality HMD system.

The very first HMD systems were very large and almost fixed in position. Recently, a much lighter HMD called *Oculus Rift* has been developed by Oculus VR (2014). Rift is effective and inexpensive for domestic users, especially gamers. The field of view (FOV) of Rift is greater than 90° horizontal and 110° vertical. We believe that in the future, the entire FOV of human eye (i.e. 170 horizontal and 135 vertical) will be covered, which will make the players have a stronger sense of immersion. Besides, it contains 3-axis accelerometers and gyroscopes for tracking the player's head movement/orientation. In addition, it has magnetometers that make the Rift able to measure the absolute head orientation tracking by canceling the

---

[2] Giulio Di Vico. (3rd March 2014). *Ivan Sutherland - Head Mounted Display* [Video file]. Retrieved from https://www.youtube.com/watch?v=NtwZXGprxag

Fig. 19 The world's first HMD system—*The sword of damocles* created by Sutherland (1968) (*left*). Head-mounted display in older days (*left*). The rift HMD developed by Oculus VR (*right*)

drifting effect by the Earth's gravitational force. Recently, SONY is developing a similar HMD product for their game consoles, called the Morpheus (Kelion 2015). In the future, these portable HMDs will bring more virtual/augmented reality applications into our daily lives.

Wearable Display

Wearable displays are much lighter and more portable than typical HMDs. They include see-through display optics, so they can also be referred to as Optical Head-Mounted Display (OHMD). OHMD adopted either the *Curved Mirror* or *Waveguide* technique. Curved Mirror technique uses a mirror to reflect the output from a tiny projector onto the glass surface. The Waveguide technique is based on various optics technologies such as polarization, total internal reflection, diffraction, or holography of light.

The Google GLASS (2014) as shown in Fig. 20 is a ubiquitous computer in the form of an eyewear, which contains a touchpad, camera, and a see-through display. The touchpad is located on the side of Google Glass, allowing users to control the device by swiping through a timeline-like interface displayed on the screen. Sliding backward shows current events, such as weather. Sliding forward shows past

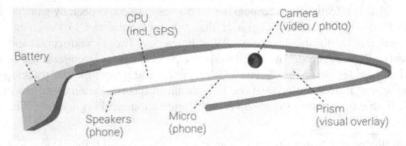

Fig. 20 The outlook of a Google GLASS. (Missfeldt 2013)

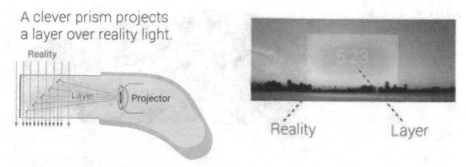

Fig. 21 The curved mirror optics technique is adopted in Google GLASS (*left*). The user's view is illustrated (*right*) (Missfeldt 2013)

events, such as phone calls, photos, circle updates, etc. The camera on GLASS is able to take photos and record 720p HD video. Users can use it to capture, search, and transmit photos and videos using apps such as messaging software.

Figure 21 (left) shows how Google GLASS adopts the Curved Mirror optics technology. The GLASS contains a mini projector which projects onto a semi-reflective mirror which only affects light stemming from the projector. The AR layer is hence capable to display on top of the reality beyond the see-through prism.

GLASS is capable to display augmented information on the see-through prism. Figure 22 shows a prototype AR application developed by researchers using NyAR-ToolKit (it is optimized and fast enough to run on portable device).It demonstrated the capability of AR running on GLASS device. However, only a picture-in-picture show can be displayed, and it is insufficient to display immersive augmented reality since the prism is too small and unable to cover a larger FOV.

Fig. 22 NyARToolkit applied in Google GLASS. (Tait 2015)

Fig. 23 The outlook of Microsoft Hololens (http://www.microsoft.com/microsoft-hololens/en-us)

HoloLens, a more immersive solution, is under development by Microsoft (http://www.microsoft.com/microsoft-hololens/en-us). It is a holographic computer that consists of a transparent lens (a see-through display device) and depth cameras. The depth information acquired reconstructs the real scene in the virtual domain, and the virtual objects can be overlaid onto the real scene beyond the lens which is aligned to the same axis references. Hence, it is able to capture the users' hand movements and allows them to pin virtual objects into the real scene like a physical object. Moreover, it is capable of playing spatial sounds that provide a more realistic presence of virtual objects Fig. 23.

Projection Techniques

Projection techniques enable the audiences to see the augmented reality with bare eyes, i.e. they need not put on any HMD or glasses. Instead, a projection media (such as a screen) is always needed.

Holography is a projection technique that enables holograms to be made, where holograms refer to 3D images. It is a waveguide optics technique that involves the usage of laser, interference, light diffraction, and light intensity under suitable illumination of the recording. The 3D image changes as the position and orientation of the viewing system changes in the same way as if the object were present in reality. Hence, the image seen by the audience appears realistic.

With the projection technique, digital avatars can be displayed "in the air" on the stage of performing arts. Hatsune Miku is a virtual singer created by Crypton Future Media (Hsu 2010). Crypton uses voices recorded by actors and runs them through Yamaha Corp.'s Vocaloid two software, in which the players can purchase the software and program Miku to perform any song on a computer. Figure 24 shows a 3D projected Miku dancing and singing on a stadium stage. In fact, "she" has been the first virtual idol in the world.

Interaction with Virtual Objects

When the player manipulates or interacts with virtual objects in the real environment, they always need a third party agent that could be either recognized by the AR toolkits or provide feedback to the system. Virtual objects and holograms could

Fig. 24 A holographic Hatsune Miku is performing her live concert on a real stage. (Wilks 2011)

appear very realistic, but they are not physical objects and hence do not have any mass. Therefore, when people press or touch them, they offer neither physical response nor reaction force. In order to give the player a stronger feeling of presence in an immersive reality-virtuality environment, simulated forces and senses have been developed.

Interaction via AR Markers

Humans cannot touch or change augmented virtual objects because they are in different space domains, even though their coordinates are overlapped perfectly with each other. An agent is required to conduct the action to change the virtual objects. As described in Sect. 2.3.1, AR markers are recognizable by camera and they can serve as a surface for displaying 3D virtual models. A virtual character can respond to the event triggered by a specialized AR marker, such as sending a gift to her (Geisha Tokyo Entertainment, Inc. 2009). It is proven that AR markers can be used as a means to interact with the augmented virtual objects.

Besides, the interactive techniques with AR markers led to useful applications. Figure 25 shows an AR furniture making tool proposed by Lau et al. (2012).The user has to put on a Head-Mounted Display (HMD) and is immersed in an augmented reality environment. The user can create 3D Furniture in the air with different AR markers, where each marker represents a particular building blockor a manipulation operation. With the HMD, the user can move around in the space and view the virtual furniture at different angles. However, the process involves many different markers and hence the user needs to learn and memorize the functionality and meaning of the markers. Hence, this kind of interaction is still not very natural.

Similarly, Tang et al. prototyped an AR application that allows user to build a complicated 3D model as shown in Fig. 26 (Tang 2015). The user can move the primitive 3D shapes in positive via moving the respective AR marker cards. The details can be edited by tapping and using simple finger gestures. Hence, the user doesn't need to memorize a lot of functions. Its interaction sounds more natural than

Fig. 25 An AR furniture making tool with HMD. (Lau et al. 2012)

Fig. 26 A 3D model built by moving the primitive shapes via AR markers. (Tang 2015)

AR furniture. However, the user needs to hold a smartphone on one hand in order to play with this app, which is not so convenient. A hands-free HMD might make this app more attractive.

Interaction with Depth Sensing AR

An additional *depth* dimension gives extra information to traditional 2D computer vision. Depth cameras have been widely used in 3D scanning (ShahramIzadi et al.

Fig. 27 The user interface of a 3D furniture arrangement system with markerless AR, making use of the microsoft KINECT depth sensor (Tang et al. 2014) (*left*). A user was operating the system (*right*)

2011) and motion capture, etc. For example, the Microsoft Kinect depth sensors (http://www.xbox.com/zh-HK/Kinect/Games) have been commercialized in domestic game and desktop markets. It is relatively cheaper than optical motion capture and high power sonar systems, and is tinier and more portable.

Researchers have prototyped markerless AR and mixed reality applications with Kinect. Tang et al. (2014) developed an application that uses the Kinect to estimate the actual dimensions of the available space in a room. Figure 27 (left) shows the user interface, which allows users to drag and drop virtual 3D furniture models into the real-time scene.Hence, the users can have a glimpse how good the furniture matches with their home environment before purchasing them. It would be great if this app could be used on a smart device with a portable depth camera.

Figure 28 shows a kind of portable structure sensor called iSense (iSenseTM 3D Scanner 2014), which is a portable infra-red depth camera. It is light weight and able to be mounted on an iPad. It is bundled with 3D scanning software that allows users to scan humans and objects and translate them into 3D models in a format ready for 3D printing. Its operation range is illustrated in Fig. 29, which is very close to the power of Microsoft Kinect.

In 2013, Toshiba has announced a new camera sensor unit TCM9518MD, which aimed to become part of future smartphones, tablets and other mobile devices

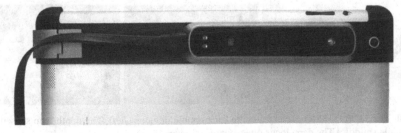

Fig. 28 The portable structure sensor mounted on an iPad (iSenseTM 3D Scanner 2014)

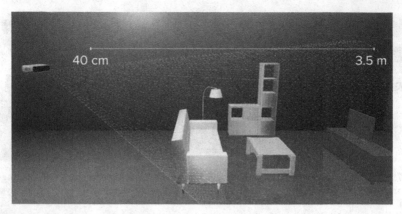

Fig. 29 The operation range of the iSense (iSenseTM 3D Scanner 2014)

(Tyson 2013). As shown in Fig. 30, the TCM9518MD camera module consists of two-one-fourth inch 5 M-pixel CMOS image sensors, which is small enough for portable devices. It uses the accompanying LSI chip to calculate depth data. One application of the depth function is producing *deep focus images* as shown in Fig. 30 (right) from the estimated depth map (middle). This depth estimation capability can be applied to refocus, size estimation, and extraction of any objects in images, and gesture operation. It makes depth sensing on smartphones possible. From the hand distance, the smartphone can estimate the world position of the user's hand, and determine whether the user is touching a virtual object. Hence, we predict in the near future the smart device (as well as wearable displays) can detect real objects and human hands. People could interact with virtual objects such as pressing a virtual button with their bare hands.

In addition to AR, depth cameras can provide a mixed reality (MR) experience to people. They scan the 3D structure of physical objects (such as the player's hands) from the reality domain, and then reconstruct and align them with the coordinates in the virtual domain. Hence, the player can touch the virtual objects directly in the air. The only difference is that the player cannot receive a feeling of touch.

Fig. 30 The TCM9518MD camera module for portable devices (*left*). The depth map sensed by the module (middle).The deep focus image (*right*) (Tyson 2013)

Fig. 31 A mixed reality character (Kim et al. 2013)

Kim et al. demonstrated this idea with *MARIO*—A Mid-air Augmented Reality Interaction with Objects system (Kim et al. 2013). As shown in Fig. 31, a virtual chicken character and its shadow are projected on different planes of a little stage. The player moves the physical blocks by hands makes the virtual chicken run around on the stage. Please find their demonstration video at the footnote[3]. Since MARIO requires no additional devices (such as wearable devices) but the player's hands, they can enjoy the natural interaction in the air.

Figure 32 illustrates the workings of the MARIO system. A depth sensor (i.e. Kinect) and a projector are hung above the table, and they are pointing vertically

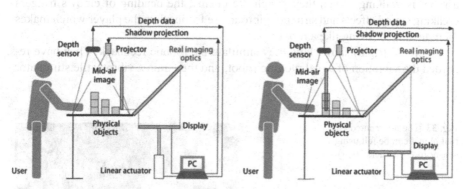

Fig. 32 The illustration of how MARIO works. When the virtual object is far away (*left*) and when the virtual object is far near (*right*) (Kim et al. 2013)

---

[3] H. Kim, I. Takahashi, H. Yamamoto, T. Kai, S.Maekawa, and T.Naemura. (2014). MARIO: Mid-air Augmented Reality Interaction with Objects [Video file]. Retrieved from https://www.youtube.com/watch?v=SpshRCmlX5Y

to the table surface. The virtual chicken on the display is projected on the imaging optics. Its shadow is visualized by the projector that indicates the depth it is located.

Feedback

Previous reality-virtuality applications could not give players a strong feeling of presence even though they are visually and auditory immersive.This is because the players are interacting with virtual objects/humans in the air without a sensation of touch or response.In order to give the players a stronger sense of presence, novel research directions such as physically feedback devices and simulated senses have been introduced.

Force Feedback

Tactile perception can be given by tangible or haptic devices. With a sensor rod, the player can touch the virtual object more intuitively and naturally. Figure 33 shows a demonstration by Harders (2013). The augmented virtual object (i.e. the white cylinder) is projected into the actual environment. The player can feel it by poking it with a sensor rod, which is a type of 3D pointing device.

Actuation of the game avatar may give the player a stronger feeling of presence. Kurihara et al. (2013) designed a robot action game that was able to provide tactile feedback. In the game, the player has to immerse his or herself by putting on an Oculus HMD. Also, they are required to put on a pair of vibrotactile gloves. Figure 34 (left) shows the game interface. When the player walks, they can feel vibration like a robot is walking. When they punch the enemy, the bending of elbows triggers creaking sound effects and stronger vibrotactile feedback to the player which makes them feel like moving like a robot.

Figure 34 (right) shows how they simulate the robotic feedback. They have recorded the vibration of a real moving robot, and then train a vibrotactile stimulator

Fig. 33 The augmented virtual object can be felt using a sensor rod. (Harders 2013)

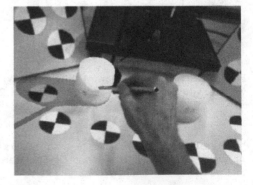

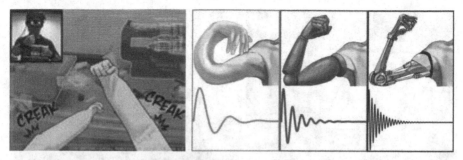

Fig. 34 The Jointonation system renders an impact vibration to the multiple joints, when the player's motion stops. (Kurihara et al. 2013)

with the data by spectral approximation analysis. Hence, the vibrotactile stimulator is able to give realistic robot-like body sensations when the player fights.

Virtual Olfaction and Gustation Senses

Virtual haptic sensation has become more prominent in both academia and in commercial applications such as game controllers (Israr et al 2012). More interestingly, digital gustatory and olfactory research is beginning to emerge that will ultimately lead to tying all of our senses together to experience new forms of entertainment previously not possible.

Figure 35 shows two laboratory prototypes that present virtual tasting experiences to the user. Food Simulator (P. Kortum 2008) uses chemical and mechanical linkages to simulate food-chewing sensations by providing flavoring chemicals, biting force, chewing sound, and vibration to the user. TasteScreen (Maynes-aminzade 2005), is attached to top of the user's computer screen and holds 20 different flavor

Fig. 35 Food simulator (*left*) and TasteScreen (*right*)

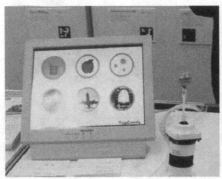

Fig. 36 Tag candy (*left*). Meta cookie (*right*)

cartridges which are mixed and sprayed on top of the display. The user enjoys the dispensed taste by licking his or her computer screen display.

Some systems have used an augmented reality approach to change the taste perception during the experience. Narumi et al. described a pseudo-gustatory display based on virtual color of a real drink (Narumi et al. 2010a). They used a wireless LED module attached to the bottom of a transparent plastic cup to superimpose a virtual color into the drink. In addition, Tag Candy (2010) used vibration and hearing through bone conductivity to deliver various sensations while a user enjoyed a lollipop attached to the system. Conversely, Meta Cookie (Narumi et al. 2010b) used visuals and smell information to provide various taste sensations to the user while consuming a real cookie Fig. 36.

A different approach is where taste is actuated using non chemical means like electricity. Alessandro Volta was one of the first scientists who studied this phenomena using two coins made of different metals and placed them on both sides of his tongue. When he connected the coins to an electric current, he recorded a salty sensation on his tongue (Volta 1800). This has found real usage in the medical field where it is known as Electrogustometry, a technique used to access the taste detection thresholds of patients with taste disabilities (Stillman et al. 2003).

Nakamura et al. showed the possibility of using this for augmented gustation (Nakamura and Miyashita 2011). They applied electric current through isotonic drinks and food to change their taste perception. Pulsing voltage and amplitude input creates variations in taste. They created several apparatus, shown in Fig. 37 above. When the user drinks through the straw or uses the chopsticks, the electric circuit is completed, and electricity is able to flow from the apparatus to the tongue and back out to the apparatus again Fig. 38.

Ranasinghe et al. (2012) developed Tongue Mounted Digital Taste Interface using silver electrodes and a peltier hot plate. The user sticks his or her tongue between the electrodes to facilitate the actuation. The peltier plate is attached to the top electrode to warm and cool the tongue, a technique known as thermal stimulation to approximate tastes like spicy and minty. Ranasinghe et al. (2014) has also developed a drinking platform called FunRasa. This uses a drinking straw outfit-

Fig. 37 Augmented drinking and chopstick apparatus

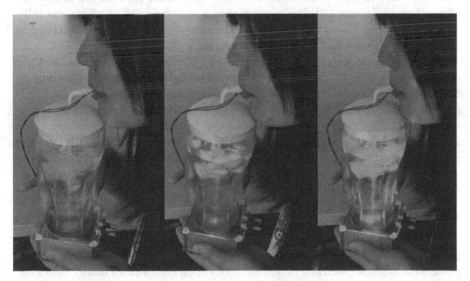

Fig. 38 FunRasa

ted with silver electrodes along with RGB LED to change the perception of the drink. The user can change the color as well as the strength of the drink using dials mounted at the bottom of the glass.

The Smell-O-Vision (Time Magazine 1959) was one of the earliest and most famous attempts to commercialize smell to augment other media like moving pictures. The system was deployed in a movie theatre where odors where released during the presentation of a movie so the viewers could smell what was being depicted.

Okada (Bannai et al. 2010) developed an "olfaction printer" from a standard ink jet printer that pulses small amounts of scent chemicals into the air. The iSmell (2012) was a conceptual prototype developed by DigiScents. Connected to a com-

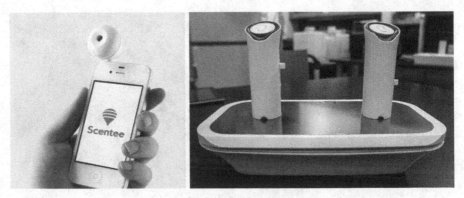

Fig. 39 Scentee (*left*) and oPhone (*right*)

puter via standard USB it would emit a scent when users browsed the internet or checked email. Trisenx (2012) also developed a similar prototype which was fully programmable and allowed users to create their own scents. It was also intended to apply to a broader array of computer applications such as games and movies.

The most recent smell products are the Scentee (https://scentee.com) and the oPhone (https://www.yahoo.com/tech/harvard-scientists-send-the-first-transatlantic-smell-89078729859.html). The Scentee connects to a smartphone via the headphone jack and emits a single scent when the user triggers an event in an app or sends a text message. The oPhone can let a user tag a photo with an "aroma note" using a proprietary smartphone app that lets them transfer the message over the internet using a smelling station on the recipient's side Fig. 39.

Conclusion

In this chapter, we have reviewed state-of-art toys and games that have been empowered with Augmented and Mixed Reality. Popular augmented reality SDKs are reviewed. Visual display is important. The trend is moving from desktop/hand-held to wearable devices. With AR technology, the rendered 3D objects are purely virtual, and they are hard to interact with. By making use of AR markers, people can manipulate the virtual object in a more direct way. Depth cameras could be useful to sense the hands' positions that aid the system in knowing which virtual object to be touched. Mixed reality (MR) and the use of Holography provide a bare-eyed reality experience. However, displayed/projected virtual objects cannot give the user a sense of touch. Additional apparatus such as haptic stylus is necessary to interact with the virtual objects. In the future, more kinds of virtual senses could be stimulated such as taste and smell.

In the future, we foresee the projection technique and wearable devices such as Google GLASS and HoloLens become more popular, and more applications based on AR toolkit will be made for them in order to achieve hands-free interaction: i.e. interacting with virtual objects with bare hands without touching the monitor or holding the device. The interaction between human (hands) and virtual objects (such as buttons)and virtual humans are most likely become popular in the future. Such interaction will be aided by the exploration of depth information in real scenes, and portable depth cameras will be very useful to estimate depth and size of real objects that are to be aligned with those in the virtual world.

References

Google 2014. Project GLASS. Google Inc. https://www.google.com/glass/start/

HoloLens, Microsoft. http://www.microsoft.com/microsoft-hololens/en-us

Milgram, P., Kishino, F. (1994). A taxonomy of mixed reality visual displays, IEICE Transactions on Information and Systems Special Issue on Networked Reality (E77D), 12, 1321–1329.

Van Dam, Andries, et al. "Immersive VR for scientific visualization: A progress report." Computer Graphics and Applications, IEEE 20.6 (2000): 26–52.

H. Regenbrecht and C. Ott and M. Wagner and T. Lum and P. Kohler and W. Wilke and E. Mueller, An Augmented Virtuality Approach to 3D Videoconferencing, Proceedings of The 2nd IEEE and ACM International Symposium on Mixed and Augmented Reality, pp 290–291, 2003

Azuma, Ronald; Balliot, Yohan; Behringer, Reinhold; Feiner, Steven; Julier, Simon; MacIntyre, Blair. Recent Advances in Augmented Reality Computers & Graphics, November 2001. MR

Hiroshi Ishii,"The tangible user interface and its evolution". Communications of the ACM, Volume 51 Issue 6, June 2008.

Cruz-Neira, Carolina, Daniel J. Sandin, and Thomas A. DeFanti. "Surround-screen projection-based virtual reality: the design and implementation of the CAVE."Proceedings of the 20th annual conference on Computer graphics and interactive techniques.ACM, 1993.

Oculus VR. (2014). Oculus Rift—Virtual Reality Headset for 3D Gaming, Oculus VR, http://www.oculusvr.com.

Kelion, Leo. (4 March 2015) "Sony's Morpheus virtual reality helmet set for 2016 launch". BBC.

James Cook, People Are Complaining About The Anime Schoolgirl Sony Used To Demo Its VR Headset Read. Business Insider (1 Sep 2014) http://www.businessinsider.com/sonys-virtual-reality-system-project-morpehus-demonstrates-summer-lesson-2014-9#ixzz3VfTnU52l.

David Kolb, Sprawling Places (2008). http://www.dkolb.org/sprawlingplaces/generalo/disconti/virtuald.html.

Game News (2013). "Oculus Rift: Developers Game Jam organized for new games projects", I Love Game Reviews www.ilovegamereviews.blogspot.hk/2013/07/oculus-rift-developers-game-jam.html.

Chan, J.C.P., Leung, H., Tang, J.K.T and Komura, T. (2011). A Virtual Reality Dance Training System Using Motion Capture Technology, IEEE Transactions on Learning Technologies, 4(2), pp 187–195.

Kinect for Windows. 2015. http://www.microsoft.com/en-us/kinectforwindows/.

Kinect Games, xbox 360 + Kinect. 2015.http://www.xbox.com/zh-HK/Kinect/Games.

Billinghurst, M., Grasset, R., and Looser, J. (2005). Designing augmented reality interfaces, ACM SIGGRAPH Computer Graphics - Learning through computer-generated visualization, 39 (1), 17–22.

AppToyz Blaster. 2014.http://www.apptoyz.com.

Getting started—ARToolworks support library. 2014. www.artoolworks.com/support/library/Getting_started.

AR Toolkit. (2002). http://www.hitl.washington.edu/artoolkit

Kato, H., Billinghurst, M. "Marker tracking and hmd calibration for a video-based augmented reality conferencing system.", In Proceedings of the 2nd IEEE and ACM International Workshop on Augmented Reality (IWAR 99), October 1999.

NyARToolkit project: http://nyatla.jp/nyartoolkit/wp/.

Qualcomm Vuforia, http://www.qualcomm.com/Vuforia.

Geisha Tokyo Entertainment, Inc. (2009) ARis, www.geishatokyo.com/jp/ar-figure/.

Lasorne, F. (2009) Augmented Reality Toy, www.tuvie.com/augmented-reality-toy-by-frantz-lasorne/.

Lowe, D. G., "Distinctive Image Features from Scale-Invariant Keypoints", International Journal of Computer Vision, 60, 2, pp. 91–110, 2004.

J. Matas, O.Chum, M.Urban, and T.Pajdla. "Robust wide baseline stereo from maximally stable extremal regions." Proc. of British Machine Vision Conference, pages 384–396, 2002.

Greg Eddington, Markerless Augmented Reality, http://users.csc.calpoly.edu/~zwood/teaching/csc572/final11/geddingt/.

Mitchlehan Media, LLC (2013), AR Flashcards, http://www.arflashcards.com.

AR Dinosaur (2014). https://play.google.com/store/apps/details?id=com.bbdco. ARDINOSAUR.

Lindsay ONeal. (7 Dec. 2011). Suwappu App Gives Toys Their Own 3-D World. Techli. Retrieved from http://techli.com/2011/12/suwappu-app-gives-toys- their-own-3-d-world/.

Mixed Reality Pong with Sphero. 2015. http://www.gosphero.com/blog/mixed-reality-pong-with-sphero/.

Ullmer, B., and Ishii, H. (2001). Emerging Frameworks for Tangible User Interfaces. In Human-Computer Interaction in the New Millenium, John M. Carroll, ed.Boston: Addison-Wesley, 2001, pp. 579–601.

Ichida H., Itoh Y., Kitamura Y., Kishino F. (2004). ActiveCube and its 3D Applications, IEEE VR 2004, Chicago, IL, USA.

SoftEther (2011), QUMARION, https://www.clip-studio.com/quma/en/.

Kidtellect. (2012). Tiggly - Interactive toys and iPad learning apps for toddlers and preschoolers, Kidtellect Inc., http://www.tiggly.com.

Disney/Pixar. (2012). AppMATes, http://www.appmatestoys.com.

Sutherland, Ivan E.1968. "A Head-Mounted Three Dimensional Display," pp. 757–764 in Proceedings of the Fall Joint Computer Conference. AFIPS Press, Montvale, N.J.

Martin Missfeldt. (February 2013). Google Glass (infographic) - How it works. Retrieved from http://www.brillen-sehhilfen.de/en/googleglass/

Tait, M. (2015). NyARToolKit on Glass, Human Interface Technology Laboratory New Zealand. http://arforglass.org/?project=nyartoolkit

Hsu, T. (2010). Japanese pop star Hatsune Miku takes the stage—as a 3-D hologram, Technology (10 Nov., 2010), The Business and Culture of our Digital Lives, from the L.A. Times, Retrieved from: http://latimesblogs.latimes.com/technology/ 2010/11/japanese-pop-star-takes-the-stage-as-a-3-d-hologram.html

Wilks, J. (2011). Hatsune Miku: Live in concert? One man's experience with a digital idol (3 Mar., 2011).TimeOut Tokyo.

Lau, M., Mitani, J., and Igarashi, T. (2012). Digital Fabrication, IEEE Computer, 45(12), pp. 76–79.

Tang J.K.T. (2015). "Augmented Blocks: A Natural 3D Model Creation Interface", Technical Report, Caritas Institute of Higher Education, Hong Kong. Retrieved from http://www.cihe.edu.hk/~jtang.

Shahram Izadi, David Kim, Otmar Hilliges, David Molyneaux, Richard Newcombe, Pushmeet Kohli, Jamie Shotton, Steve Hodges, Dustin Freeman, Andrew Davison and Andrew Fitzgibbon. 2011. KinectFusion: real-time 3D reconstruction and interaction using a moving depth camera. In Proceedings of the 24th annual ACM symposium on User interface software and technology (UIST '11). ACM, New York, NY , USA, 559–568.

Tang, J.K.T., Lau, W.M., Chan, K.K. and To, K.H. (2014) AR Interior Designer: Automatic Furniture Arrangement using Spatial and Functional Relationships. In Proceedings of VSMM 2014

International Conference. IEEE Explore Conference Publications, (978-1-4799-7227-2/14 ©2014 IEEE).

iSenseTM 3D Scanner (2014), 3D Systems, Inc., http://cubify.com/en/products/isenseorder.

Mark Tyson (2013). Toshiba announces dual-lens 'depth camera' module for mobile. (27 Sep., 2013). HEXUS.net. Retrieved from: http://hexus.net/mobile/news/ general/60665-toshiba-announces-dual-lens-depth-camera-module-mobile/.

Kim, H., Takahashi, I., Yamamoto, H., Kai, T., Maekawa, S. and Naemura, T. (2013). MARIO: Mid-Air Augmented Reality Interaction with Objects, Advances in Computer Entertainment, Lecture Notes in Computer Science, Volume 8253, pp 560–563.

Kurihara, Y., Hachisu, T., Kuchenbecker K.J., Kajimoto, H. (2013). Jointonation: Robotization of the Human Body by Vibrotactile Feedback. ACM SIGGRAPH Asia 2013 Emerging Technologies, November 19–22, 2013, Hong Kong.

Matthias Harders. (2013). Personal website - Research. Retrieved from www.vision.ee.ethz. ch/~maharder/research.html.

A. Israr, Seung-Chan Kim, J. Stec, and I. Poupyrev. "Surround haptics: tactile feedback for immersive gaming experiences," In Proceedings of the 2012 ACM annual conference extended abstracts on Human Factors in Computing Systems Extended Abstracts, pp. 1087–1090. ACM, 2012.

P. Kortum, HCI Beyond the GUI: Design for Haptic, Speech, Olfactory and Other Non-Traditional Interfaces. Burlington, MA: Morgan Kaufmann, 2008, pp. 291–306.

D. Maynes-aminzade, "Edible bits: seamless interfaces between people, data, and food," in Proceedings of the 2005 ACM Conference on Human Factors in Computing Systems, CHI 2005, Portland, OR, pp. 2207–2210, April 2–7, 2005.

Takuji Narumi, Munehiko Sato, Tomohiro Tanikawa, and Michitaka Hirose. 2010. Evaluating crosssensory perception of superimposing virtual color onto real drink: toward realization of pseudogustatory displays. In Proceedings of the 1st Augmented Human International Conference (AH '10). ACM, New York, NY, USA, Article 18, 6 pages.

"Variable candy sensations using augmented reality—TagCandy", DigiInfoTV, December 1, 2010. [Online] Avaiable: http://www.diginfo.tv. [Accessed Nov. 21, 2012].

T. Narumi, T. Kajinami, T. Tanikawa, and M. Hirose, "Meta cookie," in ACM Siggraph 2010 Emerging Technologies, Siggraph 2010, Los Angeles, CA, pp. 143, July 25–29, 2010.

A. Volta, "On the electricity excited by the mere contact of conducting substances of difference kinds," in Abstracts of the Papers Printed in the Philosophical Transactions of the Royal Society of London, vol 1, pp. 27–29. 1800.

J.A. Stillman, R.P. Morton, K.D. Hay, Z. Ahmad, and D. Goldsmith, "Electrogustometry: strength, weaknesses, and clinical evidence of stimulus boundaries," Clinical Otolaryngology & Allied Sciences. Vol. 28. no.5. pp. 406–410, October 2003.

H. Nakamura, and H. Miyashita, "Augmented gestation using electricity," in Fourth Augmented Human International Conference, AH'11. pp. 34, Tokyo, Japan, March 12–14, 2011

Ranasinghe, N., Nakatsu, R., Nii, H., &Gopalakrishnakone, P. (2012, June). Tongue mounted interface for digitally actuating the sense of taste. In Wearable Computers (ISWC), 2012 16th International Symposium on (pp. 80–87). IEEE.

Ranasinghe, N., Lee, K. Y., &Do, E. Y. L. (2014, February). FunRasa: an interactive drinking platform. In Proceedings of the 8th International Conference on Tangible, Embedded and Embodied Interaction (pp. 133–136). ACM.

"A Sock in the Nose. Review, Behind the Great Wall," Time Magazine. p. 57. Dec. 21 1959.

Y.Bannai, D.Noguchi, K.Okada, and S.Sugimoto, "Ink jet olfactory display enabling instantaneous switches of scents," Proceedings of the international conference on Multimedia, MM'10. pp. 301–310. New York. October 2010.

"How internet odors will work," HowStuffWorks, Jan. 5, 2012. [Online] Available: http://computer.howstuffworks.com. [Accessed Nov. 20, 2012].

https://scentee.com

https://www.yahoo.com/tech/harvard-scientists-send-the-first-transatlantic-smell-89078729859. html

Advanced Sound Integration for Toy-Based Computing

Bill Kapralos, Kamen Kanev and Michael Jenkin

Abstract Despite the growing awareness regarding the importance of sound in the human-computer interface and the potential interaction opportunities it can afford, sound, and spatial sound in particular, is typically ignored or neglected in interactive applications including video games and toys. Although spatialized sound can provide an added dimension for such devices, one of the reasons that it is often overlooked is the complexity involved in its generation. Spatialized sound generation is not trivial, particularly when considering mobile devices and toys with their limited computational capabilities, single miniature loudspeaker and limited battery power. This chapter provides an overview of sound and spatial sound for use in human-computer interfaces with a particular emphasis on its use in mobile devices and toys. A brief review outlining several sound-based mobile applications, toys, and spatial sound generation is provided. The problems and limitations associated with sound capture and output on mobile devices is discussed along with an overview of potential solutions to these problems. The chapter concludes with an overview of several novel applications for sound on mobile devices and toys.

Keywords Audio interaction · Spatial sound · 3D sound · Mobile devices · Virtual environments · Augmented reality · Toy

B. Kapralos (✉) · K. Kanev
Faculty of Business and Information Technology, University of Ontario Institute of Technology, Oshawa, Canada
e-mail: bill.kapralos@uoit.ca

B. Kapralos · K. Kanev · M. Jenkin
Graduate School of Informatics, Shizuoka University, Hamamatsu, Japan

M. Jenkin
Electrical Engineering and Computer Science, York University, Toronto, Canada

© Springer International Publishing Switzerland 2015   107
P. C. K. Hung (ed.), *Mobile Services for Toy Computing,* International Series on Computer Entertainment and Media Technology, DOI 10.1007/978-3-319-21323-1_6

Introduction

Our world is filled with sounds. The sounds we hear provide us with detailed information about our surroundings and can assist us in determining both the distance and direction of objects (Warren 1983). This ability is beneficial to us (and many other species) and at times, is crucial for survival. In contrast to the visual sense, we can hear a sound in the dark, fog, and snow, and our auditory system is omni-directional, allowing us to hear sounds reaching us from any position in three-dimensional space in contrast to the limited field-of-view associated with vision. Given this omni-directional aspect, hearing serves to guide the more "finely tuned" visual attention system (Shilling and Shinn-Cunningham 2002) or as Cohen and Wenzel describe (1995), "the function of the ears is to point the eyes". It has been shown that sounds can be superior to visual stimuli for gaining attention (Posner et al. 1976), and certain sounds (e.g., the sounds of a baby crying) immediately activates mental images and schemata providing an effective means of attention focus (Bernstein and Edelstein 1971). In fact, sounds not only help to focus our attention, but once the attention system is focused, sounds can help maintain our attention on appropriate information while avoiding distractions, thus engaging our interest over time (Bishop and Cates 2001). Sounds also serve to elaborate the perception of visual information by providing us with information on invisible structure, dynamic changes, and abstract concepts that may not be expressed visually (Bishop and Cates 2001).With respect to mobile services and human-computer interactions with application software, and video games, sound can play a vital role in the communication of information. Sound can also play a vital role in toys, that is, products intended for use in learning or play (although here we are particularly interested in such devices that have been augmented with computational and acoustic input/ display capabilities). In such applications, sound can be used to convey alarms, warnings, messages, and status information (such as an incoming email, or an error) (Buxton 1990). Sound effects, known as Foley sounds, associated with a particular visual imagery (such as footsteps, a door opening, glass breaking, a ball bouncing, etc.), are often added in the post-production of live action film and animation (see Doel et al. 2001) to enhance the effect of the moving imagery. It is commonly accepted within the audio/entertainment industry that "sound is emotion" and a visual interface without an appropriately designed audio component will be "emotionally flat" (Doel et al. 2001). Studies regarding the role of sound in media have shown that there is an increased physiological response in players when playing video games with sound versus those playing without sound (Shilling et al. 2002).

Despite the important role of sound within the human-computer interface, visual-based interactions comprise the majority of human-computer interactions. Recently, however, there has been a large push to exploit other modes of interaction, particularly the use of sound for human-machine interaction. This push has been motivated by a number of factors including the following (Frauenberger et al. 2004): (i) in the content of human-machine interfaces, the information being communicated is becoming more complex, making it difficult to express essential information using visual cues alone, (ii) in many applications, visual cues are restricted by the user's mobility, form factors, or by the user's visual attention being employed for other tasks, and (iii) given society's reliance on computers, such devices should be available to all members of society, including the visually impaired who cannot make use of visual-based displays. Fortunately, there is a growing effort to include those with disabilities of various types including visual impairment, in all areas of the "technological revolution" (e.g., video games). For example, audio-only video games that are played and perceived using sound, music, and acoustics only, provide access to video games by the visually impaired. One driver to this is the introduction of government regulations in many countries requiring software systems to address the specialized needs of the disabled and require systems to be accessible.

With the latest technological advances and ubiquitous use of mobile devices, sound can be used to provide unique, engaging, and interactive user experiences. Consider the Google Glass wearable computer that includes an optical head-mounted display. Google Glass provides information to the user in a hands-free format and allows users to communicate with the Internet using speech-based commands. After issuing an Internet search, for example, the speech-based search response is output to the user using bone conduction through a small loudspeaker located beside the ear, thus ensuring that the sound is (almost) inaudible to others (Arthur 2013). Sound is often included in toys, and in many video games sound is an integral part of the toy or game (e.g., the Guitar Hero series of music rhythm video games). Traditionally, toy sound was confined to the output of pre-recorded sound (sound effects and/or speech) through a single, poor quality loudspeaker (a discussion regarding loudspeaker-based sound is provided in Section "Headphones versus Loudspeaker Output") and offered little, if any, interaction possibilities. However, more recent toys and games often incorporate mobile devices and augmented reality technologies to provide far greater interactive and engaging user experiences, particularly with respect to sound. For example, the Tek Recon blaster toy (2014) features "real triggers and recoil action" and fires specially designed reusable "soft rounds" that can reach up to 23 m without endangering the public. Tek Recon has developed a freely available iOS and Android apps that make use of global positioning system (GPS) and mobile technology to provide an interface complete with interactive heads up display, live chat, radar tracking, and sound effects to accompany each firing of the blaster. The smartphone is attached to the blaster allowing for a "battle" amongst multiple players in the real world (see Tek Recon 2014).The software serves as a heads-up display allowing players to see how much ammunition (ammo) is left in their blaster, provide access to different vision modes such as night vision and a heat sensing view, and to "see" the location of enemies and teammates.

Sound-Centered Games and Toys

Sound effects and speech have played a central role in various influential toys, games, and video games over the years including the very popular Speak & Spell,

Simon, Merlin, and Operation. The Speak & Spell educational toy was introduced in 1978 by Texas Instruments as a tool for assisting children to learn to spell and pronounce commonly misspelled words. It was the first toy to incorporate electronically synthesized speech (Ostrander 2000). The user is presented with a spoken word (generated using a speech synthesizer) and their task is to correctly spell the word using the device's keyboard. When the user spells the word correctly, they are verbally praised but if they spell the word incorrectly they receive encouragement to try again. Speak & Spell supported the use of cartridges allowing for additional content to be added (words to be spelled/pronounced). Another early example of the use of sound in computer games is Simon. Simon was an assembly language-based electronic game developed by invented by Ralph H. Baer and Howard J. Morrison (1980) and released in 1978. The game device itself includes four colored buttons (green, red, blue, and yellow) each of which generates a particular tone when pressed or activated by the device (all tones are harmonic). The device automatically lights up a random sequence of buttons and the tone associated with each button. The player must then press the buttons in the same order as they were presented. After each round, the difficulty of the audio-visual pattern is increased by increasing the number random buttons presses required by the player (Baer and Morrison 1980). Some early hand-held devices also supported audio generation. Developed by Parker Brothers in 1978, Merlin is one of the earliest and most popular hand-held (touch phone-like) gaming devices. Through a series of buttons and sound output, Merlin supported six games: (i) Tic-Tac-Toe, (ii) Music Machine, (iii) Echo (a game similar to the game Simon), (iv) Blackjack 13, (v) Magic Square (a pattern-based game), and (vi) Mindbender (a game similar to the game Mastermind). Prior to the development of electronic games supporting audio generation, a number of mechanical and electro-mechanical games included sound as a key component of the game. For example, in the "Operation" game, the player takes on the role of the doctor and must make Cavity Sam (the patient) better or "get the buzzer". Cavity Sam is cured by picking funny ailment pieces out of the game tray using a pair of tweezers. However, if while attempting to remove a piece the player touches the sides of the openings with the tweezers, they will get the buzzer and light up Cavity Sam's nose; the player who removes the most ailments wins (Hasbro 2014).

Although many video games are graphic-centered, video games can also be sound-centred. A sound-centered game can be an audio-only video game (a video game that is played and perceived using sound, music, and acoustics only), or it can be an audio-based video game where visuals are included as part of the game but are not the focus (rather, sound is the focus). As described below, a number of audio-only and audio-based video games have been developed. Many 3D audio video games allow the user to explore an imaginary three-dimensional world of some form using sound. For example, Roden and Parberry (2005) present an engine for mobile game development that employs spatial sound and speech recognition. Motivation for the development of the engine hinges on the belief that generating realistic spatial sound is easier and less computationally demanding in contrast to the generation of visuals (graphics). The engine is designed for narrative-based adventure video games in which speech is used to provide information about a story or scenario. In addition to speech, the engine provides support for background music and sounds that correspond to objects or processes. The concept of a "sound stage" is used to locate the various characters in the video game: the narrator is always directly in front of the player while the non-player characters (NPCs) and other players are placed 60° to the right and left of the player. The engine provides support for the creation of "worlds" which set the context for the storyline. In each world, the developer can establish attributes (e.g., volume and position) for the sounds and apply obstruction, occlusion, or exclusion filters to them. The user can interact with the system using speech, the keyboard, or both. The authors suggest that audio-based video games be explicitly designed such that the player is immersed in an invisible world rather than be a "blind explorer in a virtual world". Another example is Papa Sangre, is an audio-based video game developed by Somethin' Else specifically for the Apple iOS platform. This game places the player—who is dead and exists in the afterlife— in Papa Sangre's palace, in complete darkness. The player's task is to save his/her love and get out of the palace while avoiding dangerous monsters. The player must navigate through the palace using only sound cues; they must determine which direction or how far/close objects are using binaural audio conveyed over headphones (Collins and Kapralos 2012).

Finally, Ranaweera et al. (2012) have developed a mobile-based (smartphone) virtual concert application whereby instruments are arranged around a virtual conductor (the user) located at the center of the arranged instruments. Using the smartphone as a simplified baton, the user is able to control the instruments in the concert. For example, selecting a specific instrument by pointing at it and tapping to select or start playing it. Upon selecting an instrument, the conductor is provided with visual cues regarding the ensemble (e.g., the instrument is "jiggled" or its components dilated and contracted, and a spotlight appears until the instrument is muted). Although not necessarily a video game, this particular application highlights the novel and engaging sound-based interactions on mobile platforms.

Augmented Reality Audio-Based Games

In addition to 3D video games, there also exist audio-based games that are designed as augmented reality games, that is, games that combine real-world elements with game worlds. In Guided By Voices, for instance, players navigate a 3D world using a wearable computer interface (Lyons et al. 2000). Guided By Voices uses a wearable computer and radio frequency-based location system to play sounds corresponding to the user's location and state. Players move around in the real world and trigger actions in the game world. These actions may be triggered on friend and foe characters and on objects that can be collected. If the player collects the necessary objects when they reach a specific location, changes in the narrative occur. The sound design was created to answer specific questions that the user may have in each space: Where am I? What is going on? How does it feel? And what happens next? The authors note that "When creating the sounds for this environment 'real' sounds were not appropriate. It is not enough to simply record a sword being drawn from a sheath. Instead, a sound effect must match the listener's mental model of what a sword should sound like. This is especially important in this and similar games that lack visual cues. It is known that if a sound effect does not match the player's mental model, no matter how 'real', he/she will not be able to recognize it" (Lyons et al. 2000). One key addition in Guided by Voices is the use of a narrator who explains what actions have taken place when the player's character dies, since it's not always clear from sound effects alone.

Ekman et al. (2005) present the design of the mobile game, The Songs of the North. Although the game includes a visual component (graphics), sound remains the primary output mode. Similar to the work described by Friberg and Gardenfors (2004), the developers argue that sound is especially suitable for mobile devices because of the limited graphical rendering capabilities of such devices. Furthermore, they argue that using sound can free the user from having to attend to the visual display of the device and enables them to engage in games in which their movement can be an important part of the gameplay. In The Songs of the North, players explore a spirit world and interact with virtual objects and characters. Players take on the role of spirit wolves looking for magical artifacts. These "artifacts" are placed within locations in the game-world, which correspond to geographical locations in the real world. In this massively multiplayer game, players interact—collaborate or fight—with non-player characters and other human players in the game. Sound is used primarily to provide information regarding the current state of the spirit world and to provide specific information about objects, characters, and actions. Based on the mobile phone's GSM positioning, when a player approaches a location containing an artifact, his/her phone plays a different sound depending on the location and the artifact in the vicinity. At this point, the player can interact with the artifact in a number of ways (e.g., collect it, or return it to the world). The selected interaction is accomplished by generating a particular tune from the phone's drum interface. The game takes the player's position as input, and outputs notification sounds through the phone, based on a player's proximity to something he/she can interact with.

Collins et al. (2010) describe a preliminary mobile audio positioning game that employs marker-based interactions. Players are assigned a particular sound to their mobile device and provided with a corresponding positional marker to wear on a hat on their head. In a room configured with cameras, the position of the players is tracked using the ARToolKit and a large screen placed on one of the walls displays the results of the positioning in a 3D representation. Players create a soundscape to match the landscape by positioning themselves in the room according to a series of directions. They may be required to continuously move throughout the room to match the positioning of a moving item in the landscape. Points are awarded for quickly locating the correct position of their sound-making object within the landscape.

Spatialized Sound and the Human-Computer Interface

Although the awareness of sound within the human-computer interface is growing, sound, and spatial sound in particular, is often ignored and/or neglected in interactive virtual environments (such as virtual simulations, video games, and serious games), and toys despite the great interaction opportunities it can provide. Spatial sound technology refers to modeling the propagation of sound while within an environment while accounting for the human listener, or as Väljamäe (2005) describes it, the goal of spatial sound rendering is to "create an impression of a sound environment surrounding a listener in 3D space, thus simulating auditory reality". Spatial sound technology goes beyond traditional stereo and surround sound by allowing a virtual sound source to have such attributes as left-right, back-forth, and up-down (Cohen and Wenzel 1995). Spatial sound within interactive virtual and augmented reality environments allows users to perceive the position of a sound source at an arbitrary position in three-dimensional space and when properly reproduced, it can deliver a very life-like sense of being remotely immersed in the presence of people, musical instruments, and environmental sounds (Algazi and Duda 2011). Spatial sound can add a new layer of realism (Antani et al. 2012), and contributes to a greater sense of "presence" (i.e., the sensation of "being there"), or "immersion" (Pulkki 2001) (see Nordahl and Nilsson (2014) for a thorough discussion on presence and the influence of sound on presence). Spatial sound can also improve task performance (Zhou et al. 2007), convey information that would otherwise be difficult to convey using other modalities (e.g., vision) (Zhou et al. 2007), and improve navigation speed and accuracy (Makino et al. 1996). When properly reproduced over headphones, spatial sound can deliver a very life-like sense of being remotely immersed in the presence of people, musical instruments, and environmental sounds whose originals are either far distant, virtual, or a mixture of local, distant, and virtual (Algazi and Duda 2011). Nordahl and Nilsson (2014) suggest that auditory stimuli should be "regarded as a necessary rather than simply a valuable component of immersive virtual reality systems intended to make individuals respond-as-if real through illusions of place and plausibility".

Despite the importance of sound in the human-computer interface and the novel interaction opportunities sound can provide, there are a number of problems and issues that must be addressed particularly with respect to spatial sound generation on the currently available computing platforms such as interactive surface computers (table-top computers, touchscreen tablets such as the Apple iPad). Not only are such computing platforms typically limited computationally (although this is improving), but they represent challenging technical issues given the assumption of such devices that the user will be positioned in front of the display and the information will be presented to them vertically, may not necessarily hold. Furthermore, mobile platforms have generally relied on a single (and often poor quality) loudspeaker to convey sound to the user greatly limiting what can be done with respect to spatial sound. However, incorporating sound within the user interface of mobile platforms/devices is important particularly given the reduced capacity inherent with the smaller visual displays associated with such devices limits the presentation of ordinary text and graphics and makes visual immersion with such devices difficult. In other words, as described by Fernando et al. (2007), the size of the visual interface is intimately related to the amount of information that can be conveyed. With its independence of the visual display, spatial sound can be used to provide a high level of auditory immersion making it a "natural choice for mobile applications" (Algazi and Duda 2005).

Fig. 1 An example of a
vertically aligned television
screen with users viewing it
directly ahead of them while
seated

Finally, for many decades now we have experienced our audio-visual media on screens that have been aligned vertically with users/viewers sitting or standing in front of the screen looking directly at it. As illustrated in Fig. 1, televisions, movie theaters screens, and computer screens have all presented information vertically in front of us and as a result, sound (music, dialogue, and sound effects) for television, film, software, and video games has been designed accordingly, with the placement of the loudspeakers and the sound mixing all developed based on this "vertical" format. However, with the recent surge in the use of mobile devices (smartphones, and tablets), and (to a lesser degree) tabletop computers (also known as surface computers, smart table computers, or smart tables), where users position themselves around a horizontal computer screen (in a manner similar to sitting around a "traditional" table; see Fig. 2), while look down at the screen, the assumption that users stand in front of the screen vertically cannot always be made. Furthermore, tabletop computers are designed as multi-user devices with viewers at several different angles and positions. In other words, no longer is there just a single user in front of a display looking at it vertically (see Fig. 2). The move from vertical screen-based

Fig. 2 Two users seated
around a tabletop computer
with a horizontally-aligned
display

digital games to a tabletop introduces interesting open questions with respect to the delivery of visuals and sound. More specifically, as outlined by Lam et al. (2015), when we move to a horizontal screen, where do we position the loudspeakers when there are two users opposite each other playing a game (i.e., where is the "front")? How does our perception of spatial sound change when we are leaning over our computer screen versus facing it? Where should we place the loudspeakers? Where should we position sounds in the mix, (in which speaker) for best reception by the participants?

Chapter Overview

In this chapter we provide an overview of sound and spatial sound for use in the human-computer interface with an emphasis on mobile devices. In particular we concentrate on audio input/display. Such audio capabilities may be part of a suite of interaction modalities, or they may be the primary mode of interaction with the device. To take just one example of the latter, Dolphin (2014) refers to a "sound toy" as a device specifically intended as a playful medium for composition that provides access to music composition and sound creation. Sound toys can be considered as "compositional systems that allow players access to parameters of composition, various types of musical experiences, and sound worlds". They describe playful, accessible, and exploratory sonic-centric audiovisual interactive composition systems and software applications where the term "toy" suggests playful interactions (Fernando 2007). We provide an overview of the problems/limitations associated with sound capture and output on mobile devices along with potential solutions to these problems. The remainder of the chapter is organized as follows. An overview of spatial sound generation (auralization) and its associated limitations/issues along with a discussion of these limitation can be potentially overcome is provided in Section "Spatial Sound Generation". Section "Sound Capture and Output" provides a discussion regarding the capture (recording) and output of sound. The focus of the discussion is on mobile devices and more specifically, the issues inherent with the miniature inexpensive loudspeaker common on mobile devices.

Spatial Sound Generation

Collectively, "the process of rendering audible, by physical or mathematical modeling, the sound field of a sound source in space, in such a way as to simulate the binaural listening experience at a given position in the modeled space" is known as auralization (Kleiner et al. 1993). The goal of auralization is to recreate a particular listening environment, taking into account the acoustics of the environment and the characteristics of the listener (a thorough review of virtual audio and auralization is provided by (Kapralos et al. (2008)). Auralization is typically accomplished by

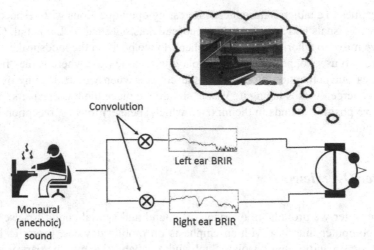

Convolution

Left ear BRIR

Monaural
(anechoic)
sound

Right ear BRIR

Fig. 3 Auralization overview

determining the binaural room impulse response (BRIR). The BRIR represents the response of a particular acoustical environment to sound energy and captures the room acoustics for a particular sound source and listener configuration. Once obtained, the BRIR can be used to filter, typically through a convolution process, the desired anechoic sound. When this filtered sound is presented to a listener the original sound environment is recreated (see Fig. 3). The BRIR can be considered as the signature of the room response for a particular sound source and human receiver. Although interlinked, for simplicity and reasons of practicality, the room response and the response of the human receiver are commonly determined separately and combined via a post-processing operation to provide an approximation to the actual BRIR (Kleiner et al. 1993). The response of the room is known as the room impulse response (RIR) and captures the reflection properties (reverberation), diffraction, refraction, sound attenuation and absorption properties of a particular room configuration (e.g., the environmental context of a listening room or the "room acoustics"). The response of the human receiver captures the direction dependent effects introduced by the listener due to the listener's physical make-up (e.g., pinna, head, shoulders, neck, and torso) and is known as the head related transfer function (HRTF). HRTFs encompass various sound localization cues including interaural time differences (ITDs), interaural level differences (ILDs), and the changes in the spectral shape of the sound reaching a listener. The HRTF modifies the spectrum and timing of sound signals reaching each ear in a location-dependent manner.

Various techniques are available for determining (measuring, calculating) both the HRTF and the RIR however, a detailed discussion of these techniques is beyond the scope of this chapter (see Kapralos et al. (2008) for greater details). The output of the techniques used to determine the HRTF and the RIR is typically a transfer function which forms the basis of a filter that can be used to modulate source sound material (e.g., anechoic or synthesized sound) via a convolution operation. When the filtered sounds are presented to the listener, in the case of HRTFs, they create the impression of a sound source located at the corresponding HRTF measurement position while when considering the RIR, recreate a particular acoustic environment. However, as with BRIR processing, convolution is a computationally expensive operation especially when considering the long filters associated with HRTFs and RIRs (filters with 512 coefficients are not uncommon) thus limiting their use in real-time applications and to "high" performance computational platforms.

Fundamental to the generation of virtual audio is the convolution operation that is typically performed in software in the time domain, a computationally intensive process. In an attempt to reduce computational requirements, a number of initiatives have investigated simplifying the HRTFs and RIRs. With respect to the HRTF, dimensionality reduction techniques such as principal components analysis, locally linear embedding, and Isomap, have been used to map high-dimensional HRTF data to a lower dimensionality and thus ease the computational requirements (e.g., see Kistler and Wightman (1992); Kapralos et al. (2008). Despite the improvements with respect to computational requirements, even dimensionality reduced HRTFs are still not applicable for real-time applications and although the amount of reduction can be increased thus improving performance, reducing the dimensionality of HRTFs too much may lead to perceptual consequences that render them impractical. Further investigations must be conducted in order to gain greater insight. With respect to the RIR, it is usually ignored altogether and approximated by simply including reverberation generated with artificial reverberation techniques instead. These techniques are not necessarily concerned with recreating the exact reflections of any sound waves in the environment. Rather, they artificially recreate reverberation by simply presenting the listener with delayed and attenuated versions of a sound source. Although these delays and attenuation factors do not necessarily reflect the physical properties of the environment being simulated, they are adjusted until a desirable effect is achieved. Given the interactive nature of video games and their need for real-time processing, when accounted for, reverberation effects in video games are typically handled using such techniques.

More recent work has seen the application of the graphics processing unit (GPU) to spatial sound rendering. The GPU is a dedicated graphics rendering device that provides a high performance, interactive 3D (visual) experience by exploiting the inherent parallelism in the feed-forward graphics pipeline (Luebeke and Humphreys 2007). For example, GPU-based methods that operate at interactive rates for acoustical occlusion (Cowan and Kapralos 2013), reverberation modeling (Rober et al. 2007), and one-dimensional convolution have been developed (Cowan and Kapralos 2013). Recent work has also taken advantage of perceptual-based rendering whereby the rendering parameters are adjusted based on the perceptual system (often vision), to limit computational processing. Both of these approaches have shown definite promise in their ability to provide computationally efficient spatial sound generation for interactive, immersive virtual environments and although greater work remains, they present a viable option for spatial sound generation on mobile devices. Greater details regarding the influence of sound over visual rendering and task performance is provided by Hulusic et al. (2012) while an overview of "crossmodal influences on visual perception" is provided by Shams and Kim (2010).

Sound Capture and Output

Headphones Versus Loudspeaker Output

With any auditory display, sound is output to the user using either headphones or loudspeakers. There are advantages and disadvantages with the use of either head-phones or loudspeakers for sound display and one or the other may produce more favorable results depending on the application. The majority of mobile devices (notebook/tablet computers, cell phones, toys, MP3 players, gaming devices, toys, etc.) will generally have a single (miniature) loudspeaker. The proliferation of these devices has also led to a growing demand of higher quality sound output/reproduc-tion, and this has presented manufacturers of these devices with increasing chal-lenges while they attempt to provide louder, higher quality sound from typically small, lightweight, and inexpensive loudspeakers (Llewellyn 2011). Being low-cost, these loudspeakers rarely allow for the ideal flat frequency response across the entire audible frequency spectrum but rather exhibit significant variations in their output level across the audible spectrum, particularly at lower frequencies (Llewellyn 2011). In addition, smaller loudspeakers are generally restrictive with respect to their overall acoustical output and this too is particularly evident at lower (bass) frequencies (Larsen and Aarts 2000). Various methods and techniques have been devised to address some of these limitations inherent in small loudspeakers. For example, Larsen and Aarts (2000) describe a method that takes advantage of several human psychoacoustic phenomena that evokes the illusion of a greater lower frequency response while the power output by the loudspeaker at lower fre-quencies remains the same or is even lowered. A detailed discussion regarding this method in addition to other methods to improve the audio quality of mobile devices loudspeakers is provided by Llewellyn (2011). Aside from the technical issues asso-ciated with miniature loudspeakers, in the presence of others (e.g., on a plane, train, public space), loudspeaker-based sound may be bothersome to others. Furthermore, a single loudspeaker, common on the majority of mobile devices, cannot convey spatial sound (a minimum of two loudspeakers is required); often, headphones must be used instead.

Sound output via headphones offers several potential advantages over loud-speaker based systems. In particular, headphones provide a high level of channel separation and this is thereby minimizing any crosstalk that arises when the signal intended for the left (or right) ear is also heard by the right (or left) ear; this is particularly problematic when employing HRTF-based spatial sound via multiple loudspeakers. Headphones also isolate the listener from external sounds and rever-beration which may be present in the environment (Gardner 1998), ensuring that the acoustics of the listening room or the listener's position in the room, do not af-fect the listener's perception (Huopaniemi 1999). Headphones are typically used to output spatial sound where the goal is to control the auditory signals arriving at the listener's ears such that these signals are perceptually equivalent to the signals the listener would receive in the environment being simulated (Ward and Elko 2000).

There are various drawbacks associated with using headphones. More specifically, after wearing headphones for an extended period of time, they can become very uncomfortable (Kyriakakis et al. 1999). Headphones may interfere with potentially important environmental sounds. Furthermore, the listener is constantly reminded that they are wearing the headphones and with respect to immersive virtual environments, this may negatively influence immersion. In addition, headphones may limit natural interactions amongst multiple users of an application.

There are a number of issues with headphones specific to the delivery of spatial sound. More specifically, sound delivered via headphones may lead to (see (Kapralos et al. 2008)) (i) ambiguous cues arising when the sound source is positioned on the median plane or directly above or below the listener, (ii) inside-the-head localization resulting in the false impression that the sound is originating from inside the listener's head (Kendall 1995), (iii) small movements of the headphones themselves, while being worn by the listener can change the position of the sound source relative to the listener, and (iv) the use of non-individualized HRTFs may be problematic.

Sound Capture with Mobile Devices

In addition to the output of sound via headphones or loudspeakers, sound can form part of the input interface, allowing users to interact with their application using sound. With recent technological advances, particularly with respect to speech synthesis, speech-to-text/text-to-speech conversion, and natural language user interfaces, sound, and speech in particular, is now a common method of interaction and widely available, included with a variety of applications/platforms. For example, Google Chrome (version 11) includes a speech-to-text feature that allows users to "talk" (issue instructions) to their Chrome browser. Microsoft's Speak is a text-to-speech converter included with Word, PowerPoint, Outlook, and OneNote, and allows for the user's typed words to be played as spoken words while Apple's Siri intelligent personal assistant (available on all iPhones, 4S or later) allows users to issue speech-based commends (e.g., send messages, schedule meetings, place phone calls, etc.). Unlike traditional speech recognition software that requires users to remember keywords and issue specific speech commands, Siri understands the user's natural speech, asking the user questions if it requires further information to complete a task.

The use of speech-based interaction techniques in addition to the growing use mobile-based conference calling and video telephony (which is growing in popularity given recent advancements in wireless technologies and mobile computing power), does require the ability of capturing the user's speech from a relatively close distance (e.g., one at about arm's length (Tashev et al. 2008)). Despite great improvements with respect to computing power, wireless technologies, and speech-based interaction methods and techniques particularly with respect to mobile platforms (phones, tablets, etc.), sound capture on mobile devices is typically limited. More specifically, such devices typically employ a single poor quality omni-directional microphone (which ideally responds equally to sounds incident from all directions) that picks up excessive ambient noise and reverberation, and has a limited range (typically under one meter). However, with respect to video-based streaming and capture, microphones should be able to capture sound at distances up to three meters (Tashev et al. 2008). As Tashev et al. (2008) describe, sound capture quality can be improved (an increase in the signal-to-noise ratio) by replacing the omni-directional microphone with a directional microphone (which will respond to sounds incoming from a specific direction(s)) and by employing multiple microphones (i.e., a microphone array) with beamforming techniques. Beamforming techniques allow a microphone array to localize sound-based events and/or be steered (focused) to a specific sound source location to capture any emanating sound from a sound source while attenuating any environmental noise that may be present (Zhang et al. 2008). Essentially, the goal of beamforming is to individually adjust the phase and/or the amplitude of the signal received at each microphone such that the combined output signal maximizes the signal-to-noise ratio (Boyce 2012a, b). Although various methods of beamforming have been developed, the simplest and perhaps most common, is *delay and sum beamforming* (Johnson and Dudegeon 1993). Consider a sound source located at some position x_s in three dimensional space. Furthermore, consider an array of M microphones (each microphone is denoted by m_i for $i = 1...M$) where each microphone is at a unique position x_i and each is in the path of the propagating waves emitted by the sound source. In general, the time taken for the propagating sound wave to reach each microphone will differ and the signals received by each of the microphones will not have the same phase. The differences in the time of arrival of the propagating wave at the sensors depend on the direction from which the wave arrives, the positions of the sensors relative to one another, and the speed of sound v_{sound}. Beamforming takes advantage of these time differences between the time of arrival of a sound at each sensor (Rabinkin 1994), and allows the array signals to be aligned after applying suitable delays to steer the array to a particular sound source direction of arrival. Beamforming consists of applying a delay and amplitude weighting to the signal received by each sensor $s_i(t)$ and then summing the resulting signals:

$$z(t) = \sum_{i=0}^{M-1} w_i s_i (t - \Delta_i),$$

where $z(t)$ is the beamformed signal at time t, w_i is the amplitude weighting, Δ_i is the delay, and M is the number of sensors (Johnson and Dudgeon 1993).

The number of microphones employed in a microphone array on a portable device is typically small (two, or three) and the spacing between the microphones is typically also limited thus limiting the utility of a microphone array. Nonetheless, improvements over a single microphone generally result. Tashev et al. (2008) introduced a microphone array geometry for portable devices. This microphone array consisted of two unidirectional microphones placed back to back facing away from each other and provided well balanced noise suppression and speech enhancement that leads to an increase in the overall perceptual sound quality.

Finally, in addition to sound capture, the microphone on a mobile device can be used to pick up sounds arising from a variety of physical phenomena. This can, for example be used to support mobile music performance whereby performers can whistle, blow, and tap their devices as a method of musical expression (Zhang et al. 2008).

Beyond Virtual and Augmented Reality and Toys

The use of sound in virtual reality (VR) and augmented reality (AR) environments and toys that we have considered so far assume direct involvement of humans and their engagement in producing and/or listening to sounds. However, there are other uses of sound that may play an important role in VR and AR environments without direct engagement of humans. In our natural environment, sound properties can be instrumental for redefining the notion of presence and engagement/involvement and this can be extended to VR and AR environments and corresponding entities. The determination of physical presence based on the Euclidean distance between two entities is often not readily applicable, given the potential complexity of the involved environments. For example, consider a typical ("traditional") lecture. Here student attendance or "presence" can be verified by the instructor simply calling out a roll call of student names and receiving oral confirmations even if students are sitting at the far end of the room. Yet, a student who is just a meter or two away from the instructor in an adjacent room may not hear the instructor and thus not be "present" even if they are physically close. In the "normal/traditional" world we deem people that can hear us, and can be heard by us, as "present" and thus available for engagement in various interactions. That is, two people are present if they are within the sound of each other's voices. It is instructive to observe that this notion of presence is different from geographical nearness. Two physically close agents may not be able to hear each other. Validating this type of presence using radio signals (e.g., by Bluetooth or Wi-Fi signals) is problematic, as these signals can pass through walls and as a result, under the right conditions, agents can communicate over considerable distances. However, presence can be validated in a straightforward manner using ultrasound signals. More specifically, lower-band ultrasound signals that are beyond the range of normal human interaction can be readily transmitted and received with standard audio equipment thus providing viable mechanism for verifying presence without interfering with traditional human-agent interactions. An appropriately equipped cell phone, for example, could both emit and receive lower-band ultrasound signals to identify the agents that are 'within the sound of my voice'. In the text that follows we discuss how a similar approach can be applied to VR and AR environments and how VR and AR toy entities could be enabled with a sense of "presence" and augmented interaction capabilities through sound enhancements.

While playing, children often treat their toys (e.g., a doll) as living entities, speaking to their toys. Similarly to the voice-based communications in the classroom example discussed above, in a perfect playground the "presence" and availability of a toy should be easy to establish just by calling it. Furthermore, we can imagine an extension to this paradigm in which suitably equipped toys that want to "play" might call out to children that are present. Indeed this concept can be taken even further by observing that the toys themselves might communicate with other toys that are present.

What are the technical and physiological limits to augmenting devices with an ultrasound channel for the determination of whom and what is present? The human audible frequency spectrum falls approximately within the range of 20 –20 kHz (Moore 1989) and as a result, sound-based interaction within the human-computer interface is carried out within this range (it should be noted there is substantive evidence that humans can perceive sounds well above this frequency range (Lenhardt et al. 1991)). Consequently, sound input and output hardware is designed to cover this range within the limitations of current technologies. As discussed earlier in this chapter, the typical miniature loudspeakers embedded in mobile devices, gadgets, and toys are not always capable of delivering high-fidelity sound, especially in the low-frequency range. Furthermore, human hearing capabilities change with age (e.g., we lose sensitivity to higher frequency sounds as we age (Brant and Fozard 1990)). Obviously, such considerations do not apply to sound communications between toys and VR/AR entities where we can potentially employ a wider sound spectrum, beyond the human frequency perceptual range (e.g., ultrasound; frequencies greater than 20 kHz). In such a case, the frequency spectrum beyond the audible human frequency spectrum (e.g., *ultrasound*) could be reserved for VR/AR and toy communications while the more limited human audible frequency spectrum can be reserved for human-computer interaction-based applications. This enables VR/AR applications and toys to communicate amongst themselves without interfering any human-computer interactions, or the environment.

What are the limits and capabilities of utilizing ultrasound in VR/AR and toy environments? Our own current research and experimental work involving the use of ultrasound for VR/AR and toy communication aims to answer this question. First it is important to recognize that the deployment of ultrasound-based communication systems requires enhancements to current toy technologies. More specifically, ultrasound extends well beyond the 20 kHz human hearing limit. However, the higher the frequency, the less likely it becomes to reliably communicate the signal through the standard human-oriented hardware (e.g., loudspeakers intended for output in the human audible frequency range). Results of our own series of informal experiments indicate that input amplitudes drop 2–3 times when the signal frequency increases beyond 17–20 kHz. A feasible compromise may be to employ the upper frequencies of the human audible frequency spectrum (e.g., above 10 kHz); this range is supported by standard human-oriented sound equipment while not directly heard by most of the people. Secondly, within a VR/AR environment, a number of entities could be simultaneously engaged in different interactions. This obviously requires a mechanism for parallel communications between such entities and broadcasting. This can be addressed by adopting a simple frequency-based sound channel management system. Although limited, it allows for at least several VR/AR entities to announce their own presence and to detect the presence of others.

Our ultrasound-based work for VR/AR and toy communication has focused on serious gaming and in particular, class/lecture and/or group oriented educational gaming activities. A typical scenario involves a group of students working together in a classroom or another designated area that is supervised by an instructor. In the learning process students employ AR technologies, for example, by installing and using designated AR applications on their mobile device (smartphone). Let's consider for example a test or a quiz where traditionally students are handed a printed copy of the test, given some time to complete the test, and then the test is collected and graded. Now, what if we want students to complete the test electronically on their mobile device (e.g., smartphone, tablet, or notebook)? The test should obviously be made accessible online but only to those students that are physically present in class. Therefore, an application is required that allows downloading and opening a given test only if the device running the application is physically present in the designated room. Functionally, this can be achieved through an ultrasound beacon (the instructor's smartphone) broadcasting the access code for the test. Note that Bluetooth and Wi-Fi are not suitable for this purpose since they easily penetrate through walls unlike ultrasound.

The unique features and advantages of the proposed ultrasound communications method are clearly seen in other applications and in particular in AR games. The Tek Recon AR game as discussed earlier in this chapter employs the built-in location capabilities of the host smartphone for proximity warnings and other interaction enhancements when players are close to each other. However, there are problems with the determination of "closeness", especially indoors where GPS signals are often too weak or even non-existent. The proposed ultrasound-based method in contrast ensures reliable proximity determination both indoors and outdoors and can significantly enhance the gaming experience. By using ultrasound for two-way communications and broadcasting, for example, AR groups of partners and enemies can be established so that the behavior of humans in combat situations could be simulated more realistically in AR.

Conclusions

Although sound is a critical cue to perceiving our environment, and despite vital role sound can play in the communication of information in the human-computer interface, it is often overlooked in interactive applications such as video games, virtual environments, and toys where emphasis is typically placed on visuals. Sound within the human-computer interface is also vital when considering the visually impaired; the majority of which rely on, and are well practiced with, the use of sound (hearing) to gather information about their surroundings. Generally, human-computer interfaces have ignored the visually impaired. For example, the majority of video games are visual based and require the use of a graphical interface to allow for interaction. This poses a problem for the visually impaired who cannot make use of visual interfaces and therefore, cannot access or have limited access to videogames and all they can offer (Braud et al. 2002). However, substantive effort is currently underway to include those with disabilities of all types including visual impairment, in all areas of the "technological revolution" (including video games). This is in part to new legislation in many countries addressing the laws of the disabled and accessibility. As a result, audio-based interfaces are the natural substitution to vision and in fact, this is often exploited in the visually impaired's interaction with user interfaces (Copper and Petri 2004).

As we have described in this chapter, aside from conveying basic information to the user (e.g., an alarm sound to indicate a device's low power level), there is a great potential for novel uses of sound on AR/VR toys, beyond the traditional use of conveying information to the user. However, there are still a number of issues and limitations that must be overcome particularly when considering sound processing (and spatial sound generation) on mobile devices. Although constantly improving, the computational power in mobile devices is still a bottleneck for many interactive spatial sound applications particularly when considering that sounds are also accompanied by detailed graphics as well. Furthermore, mobile devices are typically restricted to a single miniature (and low quality) loudspeaker limiting the sound quality (particularly when considering lower frequencies), and the generation of spatial sound which requires a minimum of two loudspeakers. Further complicating matters is the limited battery life inherent in mobile devices; sound processing in combination with graphics processing can quickly deplete battery power and to avoid dropouts, sound processing must be kept at a high priority (Collins et al. 2010; Bossart 2006). However, as technology improves these problems will become less of an issue opening the door to a wide array of immersive and engaging audio-based interfaces.

Acknowledgment This work was supported in part by the *Natural Sciences and Engineering Research Council of Canada* (NSERC), the *Social Sciences and Humanities Research Council of Canada* (SSHRC), Interactive & Multi-Modal Experience Research Syndicate (IMMERSe) initiative, and the *Canadian Network of Centres of Excellence* (NCE), Graphics, Animation, and New Media (GRAND) initiative, and a KAKENHI Grant Number 25560109 (JSPS). The support of the *Research Institute of Electronics, Shizuoka University* in the form of a Cooperative Research Project Grant is also acknowledged.

References

Warren, R. M.: Auditory perception: A new analysis and synthesis, Cambridge University Press, New York, NY. USA (1983)

Shilling, R. D., Shinn-Cunningham, B.: Virtual auditory displays, Handbook of Virtual Environment Technology. In: K. Stanney (Ed.), Lawrence Erlbaum Associates, pp. 65–92, Mahwah, NJ. USA, (2002)

Cohen, M., and Wenzel, E.: The design of multidimensional sound inter- faces, Virtual Environments and Advanced Interface Design. In: W. Barfield and T. Furness (Eds.), Oxford University Press Inc., pp. 291–346, New York, NY. USA, (1995)

Posner, M. I., Nissen, M. J., Klein, R. M.: Visual dominance: An information-processing account of its origins and significance. Psychological Review, 83(2): 157–171, (1976)

Bernstein, I. H., Edelstein, B. A.: Effects of some variations in auditory input upon visual choice reaction time. Journal of Experimental Psychology, 87(2): 241–247, (1971)

Bishop, M. J., Cates, W. M.: Theoretical foundations for sound's use in multimedia instruction to enhance learning. Educational Technology Research and Development 49(3): 5–22, (2001)

Buxton, W.: Using our ears: An introduction to the use of nonspeech audio cues. In: Proc. SPIE: Extracting Meaning from Complex Data: Processing, Display, Interaction, pp. 124–127, Santa Clara, CA. USA, (1990)

Doel, K., Kry, P. G., Pai, D. K.: Foleyautomatic: Physically-based sound effects for interactive simulation and animation. In: Proc. 28th Annual Conference on Computer Graphics and Interactive Techniques (SIGGRAPH 2001), pp. 537–544, Los Angeles, CA. USA, (2001)

Shilling, R., Zyda, M., Wardynski, C. Introducing emotion into military simulation and videogame design: America's Army: Operations and VIRTE, In: Proc. GameOn Conference, London, (2002)

Frauenberger, C., Putz, V., Höldrich, R.: Spatial auditory displays: A study on the use of virtual audio environments as interfaces for users with visual disabilities In: Proc. 7th International Conference on Digital Audio Effects (DAFx'04), Naples, Italy, (2004)

Arthur, C.: Google Glass—hands-on review. *The Guardian* (London), July 2 (2013)

Tek Recon. http://www.tekrecon.com. Retrieved May 25, 2014

Ostrander, F.: The serious business of sound for toys. Los Angeles Audio Engineering Society Chapter Meeting, April 25, (2000)

Baer, R. H., and Morrison, H. J.: Microcomputer controlled game. U.S. Patent US4207087 (A), (1980)

Hasbro.: Operation game. http://www.hasbrogames.com/en-us/product/operation-game:86309B1C-5056-9047-F544-77FCCCF4C38F. Accessed on: Friday, July 18, 2014

Roden, T., Parberry, T.: Designing a narrative-based audio only 3D game engine. In: Proc. 2005 ACM SIGCHI International Conference on Advances in Computer Entertainment Technology, pp. 274–274, Valencia, Spain, (2005)

Collins, K., Kapralos, B.: Beyond the screen: What we can learn about game design from audio-based games. In: Proc. Computer Games Multimedia and Allied Technology (CGAT 2012) Conference, Bali, Indonesia, (2012)

Ranaweera, R., Cohen, M., Endo, S.: iBaton: Conducting virtual concerts using smartphones. In: Proc. 2012 Joint International Conference on Human-Centered Computer Environments (HCCE 2012), Aizu, Japan, pp. 178–183, (2012)

Lyons, K., Gandy, M., and Starner, T. Guided by Voices: An Audio Augmented Reality System. In: Proc. International Conference on Auditory Display (ICAD 2000), Sydney, Australia, (2000)

Ekman, I., Ermi, L., Lahti, J., Nummela, J., Lankoski, P., Mayra, F.: Designing sound for a pervasive mobile game. In: Proc. 2005 ACM SIGCHI International Conference on Advances in Computer Entertainment Technology, pp. 110–116, Valencia, Spain (2005)

Friberg, J., Gärdenfors, D.: Audio games: new perspectives on game audio. In: Proc. of the ACM International Conference on Advances in Computer Entertainment Technology, pp. 148–154, Jumanji, Singapore (2004)

Collins, K., Kapralos, B., Hogue, A., Kanev, K.: An exploration of distributed mobile audio and games. In: Proc. FuturePlay 2010 International Academic Conference on the Future of Game Design and Technology, pp. 253–254,Vancouver, BC, Canada, (2010)

Väljamäe, A.: Self-motion and presence in the perceptual optimization of a multisensory virtual reality environment. Technical Report No. R037/2005, Department of Signals and Systems, Division of Communication Systems, Chalmers University of Technology, Göteborg, Sweden,(2005)

Algazi, V. R., Duda, R. O.: Headphone-based spatial sound. IEEE Signal Processing Magazine, 28(1): 33–42, (2011)

Antani, L., Chandak, A., Savioja, L., Manocha, D.: Interactive sound propagation using compact acoustic transfer operators. ACM Transactions on Graphics, 31(1): Article 7, (2012)

Pulkki, V.: Spatial sound generation and perception by amplitude panning techniques. PhD Thesis, Electrical and Communications Engineering, Helsinki University of Technology, Finland, (2001)

Nordahl, R., Nilsson, N. C.: The sound of being there: Presence and interactive audio in immersive virtual reality. In: K. Collins, B. Kapralos, and H. Tessler (Eds.), The Oxford Handbook of Interactive Audio, Oxford University Press, pp. 213–233, New York, NY, USA, (2014)

Zhou, Z. Y., Cheok, A. D., Qiu, Y., Yang, X.: The role of 3-D sound in human reaction and performance in augmented reality environments. IEEE Transactions on Systems, Man, and Cybernetics—Part A: Systems and Humans, 37(2): 262–272, (2007)

Makino, H., Ishii, I., Nakashizuka, M.: Development of navigation system for the blind using GPS and mobile phone communication. In: 18th Annual Meeting of the IEEE Engineering in Medicine and Biology Society, pp. 506–507, Amsterdam, the Netherlands, (1996)

Fernando, O. O. N., Cohen, N., Cheok, A. D.: Mobile spatial audio interfaces. In Proc. Mobile HCI '07, pp. 345–37, Singapore, (2007)

Algazi, V. R., Duda, R. O.: Immersive spatial sound for mobile multimedia. In: Proc. 7th IEEE International Symposium on Multimedia (ISM 2005), pp. 12–14, Irvine, CA, USA, (2005)

Lam, J., Kapralos, B., Collins, K., Hogue, A., Kanev, K., Jenkin, M.: Sound localization on tabletop computers: A comparison of two amplitude panning methods. ACM Computers in Entertainment (2015)

Dolphin, A.: Defining sound toys. Play as composition. In: K. Collins, B. Kapralos, and H. Tessler (Eds.), The Oxford Handbook of Interactive Audio, Oxford University Press, pp. 45–61, New York, NY, USA, (2014)

Kleiner, M., Dalenbäck, B., Svensson, P.: Auralization—an overview. Journal of the Audio Engineering Society, 41(11):861–875, (1993)

Kapralos, B., Jenkin, M., Milios, E.: Virtual audio systems. Presence: Teleoperators and Virtual Environments, 17(6): 527–549, (2008)

Kistler, D. J., Wightman, F. L.: A model of head-related transfer functions based on principle components analysis and minimum phase reconstruction. Journal of the Acoustical Society of America, 91(3): 1637–1647, (1992)

Kapralos, B., Mekuz, N., Kopinska, A., Khattak, S.: Dimensionality reduced HRTFs: a comparative study. In: Proc.2008 International Conference on Advances in Computer Entertainment Technology (ACE '08), pp. 59–62, Yokohama, Japan, (2008)

Luebke, D., Humphreys, G.: How GPUs work. IEEE Computer, 40(2): 96–100, (2007)

Cowan, B., Kapralos, B.: GPU-based real-time acoustical occlusion modeling. Virtual Reality, 14(3): 183–196, (2010)

Rober, N., Kaminski, U., Masuch, M.: Ray acoustics using computer graphics technology. In: Proc. 10th International Conference on Digital Audio Effects, Bordeaux, France, (2007)

Cowan, B., Kapralos, B.: Interactive rate virtual sound rendering engine. In: Proc. 18th IEEE International Conference on Digital Signal Processing (DSP 2013). Santorini, Greece, pp. 1–6, (2013)

Hulusic, V., Harvey, C., Debattista, K., Tsingos, N., Walker, S., Howard, D., Chalmers, A.: Acoustic rendering and auditory-visual cross-modal perception and interaction. Computer Graphics Forum, 31(1): 102–131, (2012)

Shams, L., Kim, R.: Crossmodal influences on visual perception. Physics of Life Reviews, 7(3): 295–298, (2010)

Llewellyn, W.: Audio quality improvement for mobile-device loudspeakers. Planet Analog, June 26, 2011

Larsen, E., and Aarts, R. M.: Perceiving Low Pitch through Small Loudspeakers, In: Proc. Audio Engineering Society Convention 108, paper 5151, Paris, France, (2000)

Gardner, W.: 3-D audio using loudspeakers. Norwell, MA, Kluwer Academic, (1998)

Huopaniemi, J.: Virtual acoustics and 3-D sound in multimedia signal processing. Doctoral Thesis, Faculty of Electrical and Communications Engineering, Laboratory of Acoustics and Audio Signal Processing, Helsinki University of Technology, Helsinki, Finland, (1999)

Ward, D. B., Elko, G. W.: A new robust system for 3D audio using loudspeakers. In: Proc. IEEE International Conference on Acoustics, Speech and Signal Processing (ICASSP 2000), pp. II781–II784, Istanbul, Turkey, (2000)

Kyriakakis, C., Tsakalides, P., Holman, T.: Surrounded by sound. IEEE Signal Processing Magazine, 16(1), 55–66, (1999)

Kendall, G.: A 3D sound primer: Directional hearing and stereo reproduction. Computer Music Journal, 19(4), pp. 23–46, (1995)

Tashev, I., Mihov, S., Gleghorn, T., Acero, A.: Sound capture system and spatial filter for small devices. In: Proc. Interspeech 2008 International Conference, Queensland, Australia, (2008)

Zhang, C., Florêncio, D., Ba, D. E., Zhang, A.: Maximum Likelihood Sound Source Localization and Beamforming for Directional Microphone Arrays in Distributed Meetings. IEEE Transactionson Multimedia, 10(3):5 38–548, (2008)

Boyce, K.: Generating spatial audio from portable products—Part 1: Spatial audio basics. Electronic Engineering Times, March 1, (2012)

Boyce. K.: Generating spatial audio from portable products—Part 2: Acoustic beamforming using the LM48901. Electronic Engineering Times, March 16, (2012)

Johnson, D. H., Dudgeon, D. E.: Array Signal Processing: Concepts and Techniques, Prentice Hall (1993)

Rabinkin, D. V.: Digital hardware and control for a beamforming microphone array. Master's thesis, Electrical Engineering, The State University of New Jersey, New Brunswick, NJ, USA, (1994)

Misra, A., Essl, G., Rohs, M.: Microphone as sensor in mobile phone performance. In: Proc. 8th International Conference on New Interfaces for Musical Expression, Genova, Italy, (2008)

Moore, B. C. J.: An introduction to the psychology of hearing. San Diego, CA, USA, Academic Press (1989)

Lenhardt, M. L., Skellett, R., Wang, P., Clarke, A. M.: Human ultrasonic speech perception. Science, 253(5015), 82–5, (1991)

Brant, L. J., Fozard, J. L.: Age changes in pure-tone hearing thresholds in a longitudinal study of normal human aging. Journal of the Acoustical Society of America, 88(2), pp. 813–20, (1990)

Buaud, A., Svensson, H., Archambault, D., Burger, D.: Multimedia games for visually impaired children. In: K. Miesenberger, J. Klaus, W. Zagler (Eds.): Springer lecture Notes in Computer Science, 2398, pp. 173–180 (2002)

Cooper, M., Petri, H.: Three dimensional auditory display: Issues in applications for visually impaired students. In: Proc. International Community for Auditory Display, Sydney, Australia, (2004)

Bossart, P. S.: A survey of mobile audio architecture issues. In: Proc. Audio Engineering Society (AES) 29th International Conference., Seoul, South Korea, (2006)

Augmented Reality for Mobile Service of Film-Induced Tourism App

Wei-Feng Tung

Abstract A multitude of tourists seek to travel to scenic spots where popular movies and TV dramas have been filmed. Filming tourism has become a popular selling point in the tourism industry. Scenic spots where popular movies and TV dramas have been filmed attract tourists in their multitudes. This research is to develop an augmented reality (AR) mobile service for the Google Android system that can integrate location-based service (LBS) and human-computer interaction (HCI) for a 'Film-Induced Tourism' app using 'Junaio' browser technology. With this service, users can view shooting scenes of film or dramas when they are near the same locations as detected through GPS on their mobile phone. Extending on previous works, this paper puts forward AR techniques that ensure interactivity in real time, and registered in 3D spaces. Users also can search for other information such as scenic spots or restaurants near the locations of shooting spots through the AR mobile service. Additionally, in order to extend AR technology for the Film-Induced Tourism App, this study proposes a mobile service concept of interactive AR Game that can provide a further immersive experience App or Web game.

Keywords AR · LBS · 3D · App · Augmented reality · Film-Induced tourism · Junaio

Introduction

Nowadays, Augmented Reality (AR) has become an innovative mobile service on smartphones, with timely integration of digital information with multimedia elements taken from the user's environment (http://whatis.techtarget.com/definition/augmented-reality-AR). AR is a technology to combine reality and virtuality applied to the human-computer interaction extensively. AR can be a live, direct or indirect, view of a physical, real-world environment whose elements are augmented by computer-generated sensory input such as sound, video, graphics or GPS data.

W.-F. Tung (✉)
Department of Information Management, No. 510, Zhongzheng Rd., Xinzhuang Dist.,
New Taipei City 24205, Taiwan
e-mail: 076144@mail.fju.edu.tw

© Springer International Publishing Switzerland 2015
P. C. K. Hung (ed.), *Mobile Services for Toy Computing,* International Series on
Computer Entertainment and Media Technology, DOI 10.1007/978-3-319-21323-1_7

AR show real time display and visualization of several layers of information taken from the AR user's environment having diverse digital representation forms including text, image, and multimedia (http://www.digitaltrends.com/mobile/what-is-augmented-reality-iphone-apps-games-flash-yelp-android-ar-software-and-more). By reflecting the real world environment on the screen through the digital camera on the smart phone, AR displays the reality and virtual reality which can be useful to find destination or building information. However, 3D technology can be used to develop wide-ranging applications for film, television, and various Internet media. Thus, this research uses AR technology to propose a mobile service and prototype called the 'Film-Induced Tourism' App that can display some specific shooting scenes of movies or TV dramas.

The advanced Film-based character storyline AR games can be integrated into the App to be an immersive AR game based on the shooting scenes on smartphones. The research mainly utilizes 'Junaio' to develop an AR-based mobile service. Junaio is an advanced mobile augmented reality browser. It's a free, fast and easy way to enhance your world with Augmented Reality experiences based on XML and HTML5, offering Location Based services as well as Image and 3D tracking. Furthermore, this research proposes an immersive mobile service concept that can integrate AR games for the AR applications.

The rest of this work includes literature review, AR Mobile Service—Film-Induced Tourism App, and conclusion.

Literature Review

The goal of this research is to develop an AR mobile service with Tourism service and entertainment based on location based service (LBS) and Augmented Reality (AR).

Location Based Service (LBS)

A location service can be defined as technology that integrates a mobile device's location or position with other information so as to provide added value to a user (Schiller and Voisard 2004). Location-based services can be query-based and provide the end user with useful information, or they can be push-based and deliver coupons or other marketing information to customers who are in a specific geographical area (Jiang and Yao 2006).

LBS applications can use the geolocation functionality of a mobile phone or smart phone to provide people with information and entertainment. LBS can be used in a variety of contexts, such as health, indoor object search, entertainment, work, and personal life, etc. For example, Lyu et al. (2005) propose how to build an enabling Environment called ARCADE for supporting the creation of

Entertainment applications of Augmented Reality and to develop two new video objects tracking algorithms for 3D surface markers tracking and human head and face tracking (Lyu et al. 2005). As LBS are largely dependent on the mobile user's location, the primary objective of the service provider's system is to determine where the user is. The study is to propose 'LBS' that integrates AR technology and travel mobile services.

Augmented Reality (AR)

Azuma and Baillot (1997) defines Augmented Reality (AR) is a variation of virtual environments (VE), or Virtual Reality as it is more commonly called. VE technologies completely immerse a user inside a synthetic environment. While immersed, the user cannot see the real world around him (Wrox 2011). AR allows the user to see the real world, with virtual objects superimposed upon or composited with the real world. AR supplements reality, rather than completely replacing it. A third concept to highlight here is Mixed Reality (MR), that containing Augmented Virtuality (AV) and Augmented Reality (AR) between a real environment and virtual environment (Fig. 1) (Lashkari et al. 2010)

Ideally, it would appear to the user that the virtual and real objects coexisted in the same space, similar to the effects achieved in the film "Who Framed Roger Rabbit?" (Fig. 2). AR can be thought of as the "middle ground" between VE (completely synthetic) and telepresence (completely real).

Some researchers define AR in a way that requires the use of Head-Mounted Displays (HMD) (http://augreality.pbworks.com/w/page/9469035/Definition%20 and%20key%20information%20on%20AR). In order to avoid limiting AR to specific technologies, Azuma and Baillot (1997) also defines AR as systems that have the following three characteristics:

1. Combines real and virtual;
2. Interactive in real time;
3. Registered in 3-D.

2-D virtual overlays on top of live video can be done at interactive rates, but the overlays are not combined with the real world in 3-D. For example, "Jurassic Park" features photorealistic virtual objects seamlessly blended with a real environment in 3-D, but they are not interactive media. Furthermore, AR enhances a user's perception of and interaction with the real world. The virtual objects display information

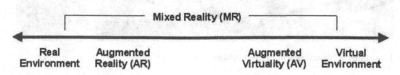

Fig. 1 Virtual continuum

Fig. 2 AR's movie 'who framed roger rabbit?'

that the user cannot directly detect with his senses, and the information conveyed by the virtual objects helps a user perform real-world tasks (Azuma and Baillot 1997). A visual AR system enhances or augments the surroundings of the user with virtual information that is registered in 3D space and seems to co-exist with the real world (Azuma et al. 2001).

AR had consolidated as technology, incorporating complex applications and systems in more and more fields of the global economy; the rapid evolution of mobile devices and the virtualization of the digital environment have prepared an auspicious environment for massive implementation of AR solutions at global level (Tutunea 2013).

Two types of AR are identified by Milgram in 1994: marker-based and marker-less. Marker-based AR implementations are used by a starting image that triggers an action when read and recognized by a digital device (phocamera, mobile device, etc.). Markerless augmented reality is more complex, extending the capabilities of a digital device with location-based functionalities (e.g., GPS location). The two types are also mentioned as location-based or position-based AR. Markerless is difficult to develop because of the technological limitations of the infrastructure (i.e. accuracy of geo-locations on GPS devices, bandwidths, respectively connection and transfer speed of technical support for communication, etc.) (Fig. 3) (Farhat and Remi 2013; Tutunea 2013)

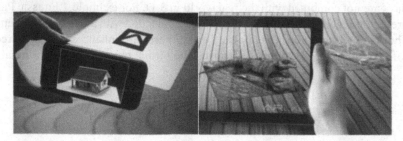

Fig. 3 Examples of visual marker and markless AR

AR Mobile Service for Tourism

A considerable number of mobile AR applications have been developed for tourism, being evidently dependent on the operating systems of mobile devices used in AR technologies, and many of the applications, even those of geo-location (Kounavis Chris et al. 2012) AR relying on location-based services make possible an extension of the functionalities offered by mobile social networks and mobile applications for tourism (Farhat and Remi 2013).

Lim et al. (2011) indicate that they combine AR with tourism to develop a tourism information system. The system can provide various kinds of information for restaurants, tourism places, regional contents, Olle trails, and historic sites in Jeju Island (Lim et al. 2011). Eguma et al. (2013) propose a system that stimulates rediscovery of sightseeing spots through hide-and-seek with CG characters using augmented reality technology (Eguma 2013). Lashkari et al. (2010) present how to create a tourist catalogue that uses AR. The catalogue will help tourists both domestic and foreign get visual aid for viewing objects on a simple map (Lashkari et al. 2010). Wei et al. (2014) indicate that two non-visual interaction modalities, haptic display and audio display, and their combination are evaluated in representing tourism information to users with a mobile phone (Wei et al. 2014).

AR Games

In recent years, AR games have developed in the game industry and presented a solution based on using Open GL 1.1 and the Android SDK for developing a multiplayer type action game belonging to the location-based mobile AR category (Jacob et al. 2012). In a typical GPS-based smartphone AR application for outdoor use, the user points the device towards physical objects in her surroundings (Yovcheva et al. 2012). Some Apps are to provide a combination of tourisms and scenes of movies.

Virtual reality (VR) generally uses 2D/3D modeling to emulate people (role), object, and scene to close the effects of reality (http://www.mortonheilig.com/InventorVR.html). In augmented reality (AR), people and scenes can be real objects, photographs, videos, or live videos as the backgrounds or scenes. AR needs to create objects or models when their interactions happen in the areas of people-objects or people-machines. AR applications just need to deal with less image processing than VR, and are appropriate for AR games, tourism, E-commerce, education, military science, design and manufacturing, social networks, and agriculture (Xiao-Jun et al. 2013) (Fig. 4)

In addition, Niantic Labs of Google provides an AR game called 'Ingress'. In this game, the participants can have different interactions with real objectives in the real world. For example, one shop can be a game's goal setting and their specific products can be the goals of game. The Ingress platform can conduct these games to provide users the ability to participate in developing various AR games. In the future, enterprises can utilize the 'tasks' to design the AR game for customers' participation and enhance their business opportunities.

Fig. 4 AR for tourism, education, translation, and interactive games

AR Mobile Service—Film-Induced Tourism App

This research is to propose an AR Film-Induced tourism (FIT) App that can provide a combination of film shooting scenes and scenic spot of tourism using AR technology and location-based GPS. Beeton (2005) indicates that film-induced tourism can be a "visitation to sites where movies and TV programs have been filmed as well as to tours to production studios, including film-related theme park". AR is an enhanced version of reality created by the use of technology to overlay digital information on an image of something being viewed through a device. With respect to games or toys, this allows the user to enhance the physical world as they play. AR can provide a way to interact in an image top of another between a virtual world and the real world. AR has two types of image identification as follows (http://whatis. techtarget.com/definition/augmented-reality-AR).

1. **Marker-based Tracking**: A webcam can catch the images and conduct marker detection. According to the marker of quadrangle, the form change of 2D images in the quadrangle can deduct the 3D plane of the marker and 3D transform. Then, the 3D transform can catch images again to draw 3D objects. There are often seen types of image identification such as Frame markers, Dot markers, Split markers, Data Matrix markers, ID markers, ID markers, and Template markers. These are well-developed identification methodologies of computer vision. The above types of image/vision identification also can be called marker augmented reality (Marker AR).
2. **Natural-Feature Tracking**: This is the first natural tracking with free and real tracking for hand held AR, and it is also referred to as Markerless AR.

The natural-feature tracking can also be used with other identification technology. Natural-feature tracking can combine frame markers to track real objects together. In terms of pattern recognition, Natural-Feature Tracking is different with the marker-based tracking and needs to set the links of the objects and graphics for recognizing. The tracking technology is able to compromise virtual objects and real objects.

Junaio

Junaio is an advanced augmented reality browser that can view thousands of channels offering digital information and contents connected to real products, locations, newspapers or billboards. However, our AR application needs to support the photos of life, real spots, and live videos to achieve the objective of combination of reality and virtual reality rather than 2D/3D modeling of virtual reality. For this reason, we employ an AR development tool called 'Junaio' to deploy a system prototype of AR apps created for the use of GPS data from Google Latitude, Google Maps, videos, sounds, and graphics to overlay digital information on images or videos being viewed through an Android mobile device. Junaio is an entry point to developing and publishing augmented reality and location-based experiences, and uses AREL (Augmented Reality Experience Language) to define its content. Junaio not only supports location-based augmented reality, but also image-based tracking, so you can focus on creating the best experience. As a free app for Android and iOS devices, Junaio has several million users and a thriving international and professional developer community (http://www.junaio.com/).

Junaio provides a mobile AR platform that allows users to unlock digital information from places, products and sights around them. Using AR technology, Junaio visualizes information in a completely new way. Junaio is powered by Metaio, a pioneer and leader in Augmented Reality technology. Now a new Beta version, Junaio Mirage, is available for smart glasses (Epson, Vuzix, Google) (http://www.junaio.com/).

Users can use a digital camera to focus any object on which they want to add information. This is a function of 'Scan the World' that can identify figures, position, QR code, or product bar code. The technology can search for the periphery information conveniently, and requires IOS4.3 or Android OS 2.2., as well as provide API and public channel for developers to locate and identify images. A Channel is a way to develop the application procedures in Junaio, where a user can develop a channel to view information. A channel can be divided into Location based channels, Junaio GLUE channels, and 3D components. Junaio can provide mark tracking, image tacking, 3D object tracking, AR gravity induction, LBS services, indoor navigation, simultaneous location and mapping (SLAM), and continue vision. This study uses Janaio browser, open API, LBS information, and an AR game to develop an AR platform for film-Induced App. (Fig. 5)

Film-Induced App

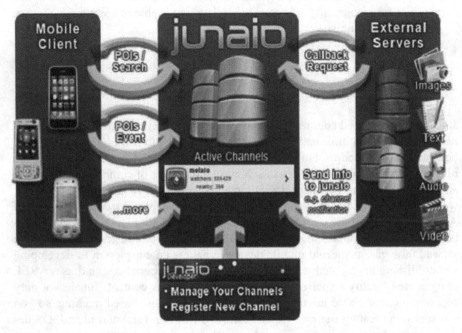

Fig. 5 Junaio framework

In order to design and develop the proposed Film-Induced App, this research utilizes Google Android and Junaio as part of the software framework and deployment of devices which are implemented for this Android App (Fig. 6).

Our Android app is designed in Java and the Android SDK development tool, as well as the API on an Android development platform. Mobile marketing material can be designed with certain trigger images that, when scanned by an AR enabled device, use the image. The innovative mobile service with HCI application for filming location tours or themed trips is able to achieve the objectives of increasing business opportunities and market benefits of peripheral products or services.

The proposed LBS's mobile phone app provides the two options for users (movies or TV dramas). The mobile service displays the map and spots' tags on the users' smartphone automatically when the digital camera is used to take a photograph. These UI of AR mobile services are shown in Fig. 7.

A main AR mobile service is able to provide the presence of shooting scenes when user is at the same address according to the map locations. An immersive shooting scene can be displayed on the smartphone when the mobile service reads the specific coordinates from GPS. For example, Fig. 8 shows that a shooting scene

Node No.	Name	Devices/Software
1	Server	Windows 7 Eclipse & SQL server
2	User Mobile Phone	Google Android Junaio
3	Manager Computer	Windows 7 Eclipse Dreamweaver Adobe Flash
4	User Computer	Browser, OS

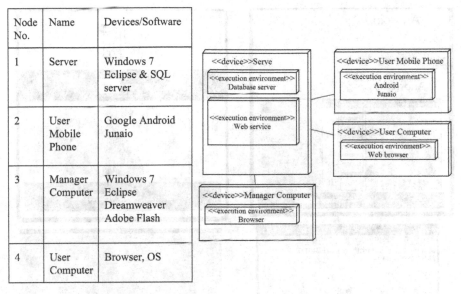

Fig. 6 Software framework and deployment

of a local TV drama is displayed on Android mobile phone through the AR mobile service.

In addition to real time display of the shooting scene on the smartphone, the other amazing interactive AR technology can be further integrated into an AR game which allows users to have an immersive experience in the shooting scenes. For instance, the Hollywood picture 'Avatar' and Coke Zero provided an interactive AR game in 2009 (https://www.youtube.com/watch?v=APQ2OxgCNzE). Coke Zero brings people access to the world of Pandora in which people entered the 3D shooting scenes of Avatar. 3D Avatar shots were taken in Zhangjiajie, China. Avatar and Coca-Cola Zero have been delivered a Web-based application enabling users to interact with 3D motion graphics—Coke Zero drinkers can engage with the movie on a uniquely visceral level. Avatar provides a unique immersive experience for moviegoers, and this promotion with Coca-Cola Zero will bring fans even deeper into the amazing world of Pandora. By holding an Avatar-branded Coke Zero can before a Webcam, visiting AVTR.com or taking a picture of the activating AVTR mark or Coke Zero logo with certain camera phones, consumers are able to access the technology, which allows them to use a computer keyboard to trigger actions such as shooting missiles or flying helicopters and firing their guns (Fig. 9).

Through AR technology, people will be immersed in an authentic and exclusive scenario. This research has developed an AR Film-Induced App using Junaio, which can show movie/TV drama pictures or clips shooting scenes when user is in the same place. AR games like Avatar and Coca-cola Zero can be referred to in future research to strengthen our Film-Induced app in order to create a much more interactive experience.

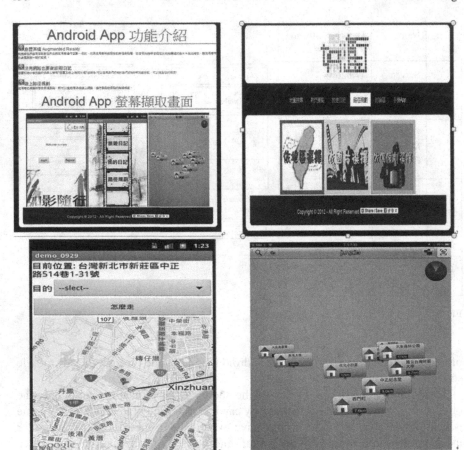

Fig. 7 UI of film-induced App

Conclusions

As people usually have trips akin to pilgrimages to famous filming sites for decades, this research presents an AR Film-Induced mobile service to increase the interest of tourists. The main objective of this research is to propose an AR mobile service for combining film and tourism for viewing the shooting scenes on a journey. AR technology can support mobile on-site needs of tourists on their smartphones through the following three functions: (1) provide access to location-based information relevant to the immediate surroundings of tourists, (2) enable access to variable timely and updated content, (3) offer interactive pictures/films of shooting scenes using LBS and GPS.

For research environment and AR solution developers, the results of this study can be shown in an AR App designed to increase value-added tourism. In addition to its function as a type of tourism tool, the App can also create marketing opportunities for peripheral shops and businesses. Since tourists visiting such spots are

Fig. 8 AR mobile service of film-induced APP

Fig. 9 AR game of Avatar and Coca-Cola zero

most likely to be fans of those films/TV dramas, the commercial opportunity of peripheral shops also can be promoted indirectly. For AR games, the proposed App can be further strengthened as an interactive 3D game which immerses the user in the shooting scenes of films or TV dramas. The 3D interactive AR game can be integrated into new applications and devices, which can be called Film-Tourism-Game AR games.

References

http://www.junaio.com/

http://augreality.pbworks.com/w/page/9469035/Definition%20and%20key%20information%20on%20AR

http://researchguides.dartmouth.edu/content.php?pid=227212&sid=1891183

http://whatis.techtarget.com/definition/augmented-reality-AR

http://www.digitaltrends.com/mobile/what-is-augmented-reality-iphone-apps-games-flash-yelp-android-ar-software-and-more

http://www.mortonheilig.com/InventorVR.html

https://www.youtube.com/watch?v=APQ2OxgCNzE

Azuma, R., Baillot, A survey of augmented reality, Teleoperators and Virtual Environments, vol. 6, pp. 355–385, 1997.

Azuma, R., Baillot, Y., Behringer, R, Feiner, S, Julier, S. & MacIntyre, B., (2001). Recent advances in Augmented Reality. IEEE Computer Graphics & Applications, 21(6), 34–47.

Farhat, Tariga Avinanta, Senjay Remi (2013), Design An AR Application in Finding Perferred Dining Place with Social network Capability, Advanced Computing: An International Journal, Vol. 4, No. 4, p. 1–16.

Kounavis Chris D., Kasimati Anna E., Efpraxia D. Zamani (2012), Enhancing the Tourism Experience through Mobile Augmented Reality: Challenges and Prospects, International Journal of Engineering Business management, Vol. 4, Special Issue, p. 1–6.

Li Xiao-Jun, Xie Bo, Ye Feng (2013) 'Research and Application of Online Product Display Technology Based on Augmented Reality' Information Technology Journal, 12 (6), p. 1134–1142.

B. Jiang and X. Yao, (2006) 'Location-based services and gis in perspective', Computers, Environment and Urban Systems, Vol. 30, p. 712–725.

Beeton, S., (2005) 'Film-induced Tourism', Channel View Publications, Clevedon.

Wrox, L. Madden, 'Professional Augmented Reality Browsers for Smartphones'. Wiley Publishing, 2011.

Lyu, M. R., King, L., Wong, T. T., Yau, E., & Chan, P.W. (2005) 'ARCADE: Augmented reality computing arena for digital entertainment' In Proceeding of IEEE Aerospace conference, Big Sky, MT: IEEE.

Youngjae Lim, Yojong Park, Joenghyun Heo, Jingeol Yang, Minji Kang and Yung-Cheol byun, 'A smart phone application based on AR Jeju tourism' 2011 First ACIS/JUN International Conference on Computers, Networks, Systems, and Industrial Engineering. 2011.

Hirofumi Eguma, Tomoko Izumi, Yoshio Nakatani, 'A Tourist Navigation System in which a historical Character Guides to Related Spots by Hide-and-Seek' Conference on Technologies and Applications of Artificial Intelligence. 2013.

Arash Habibi Lashkari, Behrang Parhizkar, 'Mohamed Abdulkarim Mohamedali, Augmented Reality Tourist Catalogue Using Mobile Technology' Second International Conference on Computer Research and Development. 2010.

Side Wei, Gang Ren, Eamonn O'Neill, 'Haptic and Audio Displays for Augmented Reality Tourism Applications' IEEE Haptics Symposium 2014, 23–26 Februar, Houstaon, Tx, USA.

Joao, Jacob, Hugo, D. Silva, Antonio, Coelho, Rui, Rodrigues (2012), Towards Location-based Augmented Reality games, Procedia Computer Science 15, P. 318–319.

Yovcheva, Zornitza, Buhalis, D., Gatzidis, C., 'Overview of Smartphone Augmented Reality Applications for Tourism', e-Review of Tourism Research (eRTR), Vol. 10, No. 2, 2012.

Tutunea, F. Mihaela, (2013) 'Augmented Reality—State of Knowledge, Use, and Experimentation' The USV Annals of Economics and Public Administration, Vol. 13, Issue 2(18) p. 215–2227.

Schiller, Jochen, Voisard, Agnes (2004) 'Location-Based Services', Schiller & Voisard.

Privacy Requirements in Toy Computing

Laura Rafferty, Marcelo Fantinato and Patrick C. K. Hung

Abstract This chapter outlines the privacy requirements for a toy computing environment. The unique architecture of toy computing requires consideration of several different factors. In this chapter we investigate the privacy requirements through formal threat modeling techniques to help the reader to get more comfortable with the toy computing architecture and how it maps to privacy threats. Next, we identify privacy requirements at legislative level, identifying privacy laws and regulations which apply to this context. The toy industry has also issued regulations for toy safety; however these regulations have no mention of privacy. While parents aim to protect the privacy of their children, we investigate the unique requirements of end users. Lastly, a demo is presented as an interface for parents to configure privacy settings for their children using mobile toy computing apps.

Keywords Toy computing · Mobile services · Privacy

Section 1. Privacy Issues in Toy Computing

Table 1 outlines a comparison between a traditional toy, electronic toy, and toy computing. This illustrates how toy computing has evolved into a new paradigm, which inspires unique privacy concerns for children. Traditionally, toys have been entirely autonomous and do not have any processing or networking capabilities to communicate with any other device. While a child user is engaged with a traditional toy, it will collect and store no personal data, and require no reason for concern for a child's privacy. With the introduction of electronic toys with embedded systems,

L. Rafferty (✉) · P. C. K. Hung
Faculty of Business and IT, University of Ontario Institute of Technology, Oshawa, Canada
e-mail: Laura.Rafferty@uoit.ca

P. C. K. Hung
e-mail: Patrick.Hung@uoit.ca

M. Fantinato
School of Arts, Sciences and Humanities, University of São Paulo, São Paulo, Brazil
e-mail: m.fantinato@usp.br

© Springer International Publishing Switzerland 2015   141
P. C. K. Hung (ed.), *Mobile Services for Toy Computing*, International Series on
Computer Entertainment and Media Technology, DOI 10.1007/978-3-319-21323-1_8

Table 1 Comparison of traditional toys, electronic toys, and toy computing

	Traditional toy	Electronic toy	Toy computing
Interaction medium	Physical	Physical (buttons)	Physical: touch
	Mechanical	Sensors—e.g. light, motion	Visual: camera
			Auditory: microphone
			Sensors: GPS, motion sensors, etc.
			Wireless interface (network)
Data collection	None	Limited	High—pervasive
Data sharing	N/A	Limited or none	Many recipients
Potential to collect location data	No	Maybe	Yes
Processing capabilities	N/A	Yes—limited	Yes—advanced
Networking capabilities	N/A	Limited or none	Communicates with other devices and services
			Wi-Fi, Bluetooth, NFC, RFID, USB
Data storage	N/A	Limited to device	On device (flash memory, SD card)
			External to device (cloud, database, server)
Architecture and TCB	Autonomous	Autonomous	BYOD—device is untrusted
	Trusted	Trusted	
Platform	Closed	Closed	Open

electronic toys can have sensory capabilities, and the ability to collect and store inputted data based on the user's interactions. This data is limited and used only for the interaction, often discarded immediately. While an electronic toy has the potential to collect and store user data, it operates on an entirely autonomous platform as a Trusted Computing Base (TCB). An electronic toy has limited or no networking capability. Thus, privacy concerns are limited to nonexistent in this architecture. Toy computing inherits the privacy concerns associated with mobile devices and BYOD. While toy computing technology allocates computing power to a mobile device, this is outside of the TCB and the device is untrusted. A mobile device also has the capability to collect a wide range of context information on the user, including their location data. Toy computing architecture allows and often requires information to be shared to services and other users. One of the most prominent concerns to privacy for toy computing compared to traditional and electronic toys is the networking capability which allows for the possibility of sharing information over a network. A mobile service is able to connect through a network to many other entities, including other mobile and Web services, servers, devices, and other users. While the mobile service has this ability to connect to and communicate with an extensive and possibly unknown amount of external entities, the issue of data sharing becomes a concern.

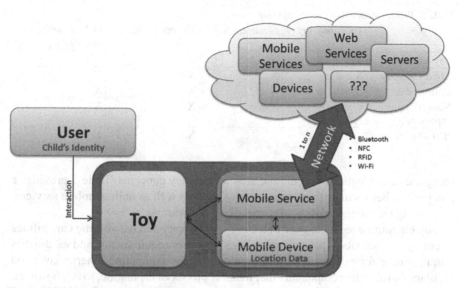

Fig. 1 Child identity and external parties

When a child engages in toy computing activities, their identity is associated with the data collected from the device. The mobile service connects to other entities over a network and shares the data. This reveals the unique privacy issues of toy computing: when a child engages with a toy computing toy, the child's identity is associated with their location data and can be shared over a network. This unique threat architecture is illustrated in Fig. 1.

These threats can be summarized as the following three items that affect the privacy in toy computing.

1. Child's Identity
2. Location Data
3. Networking Capabilities

This sharing of sensitive location data opens up vulnerabilities such as customer profiling of minors, and child predators. Customer profiling of minors involves accessing collected data to create portfolios of users related to their location history (e.g. the ability to collect data on when and where a user was, including travel patterns). This is typically used for online marketing, however location data allows for tracking and inferences on user behavior that is not otherwise publically available. Further, when physical location data is being shared with other users, it is important to consider a child's physical safety in regards to child predators who could potentially locate them based on learning their GPS location, possibly paired with other sensitive data.

While these privacy issues are common in the domains of mobile and online privacy, they are relatively new to the domain of toys. Due to the mainly child user base, and the physical toy component, toy computing separates itself from the other

Table 2 Privacy concerns in toy computing

	Traditional toy	Electronic toy	Toy computing	Online/mobile services/apps
Physical toy	X	X	X	
Child's identity	X	X	X	X
Collects data		X	X	X
Networking capability		Maybe	X	X
BYOD model			X	X

categories and identifies as a unique area for privacy concerns. Table 2 provides a comparison between the three categories of toys, as well as online/mobile services, illustrating the unique privacy concerns of toy computing.

Toy computing technology which embraces sensory and networking capabilities opens up new threats to privacy, stimulates new user requirements, and establishes a unique case for existing laws and regulations. Toy computing inherits laws and regulations from the components that make it up (services, mobile, toys), however, there are no laws that explicitly regulate the unique environment of toy computing. There is also no widely adopted framework to allow parents to manage the privacy of their children using toy computing technology. For this reason it is necessary to outline the privacy requirements to present a solution to managing location privacy for a toy computing environment.

Section 2. Privacy Threat Model

In this section, we investigate the privacy of toy computing from a threat modeling perspective. Threat modeling is a useful tool to assess risk associated with a system and provides a structured approach to security and privacy. Threat modeling can be included as part of the Software Development Lifecycle (SDL). In this section, we aim to identify location privacy threats in a toy computing environment. We present a privacy threat model for toy computing with a focus on location privacy.

Threat Modeling Techniques

Several approaches have been developed for threat modeling, one of the most widely adapted being Microsoft's Threat Modeling Process (Meier et al. 2003) (illustrated in Fig. 2) and STRIDE Model (Hernan et al. 2006) for identifying six categories of security threats: Spoofing, Tampering, Repudiation, Information Disclosure, Denial of Service, and Elevation of Privilege. This model presents an excellent approach to understanding and decomposing an application to identify security threats, however there is little focus on privacy. In order to preserve privacy, there must be a foundation of security. To achieve this, it must be ensured that the system has a reasonable

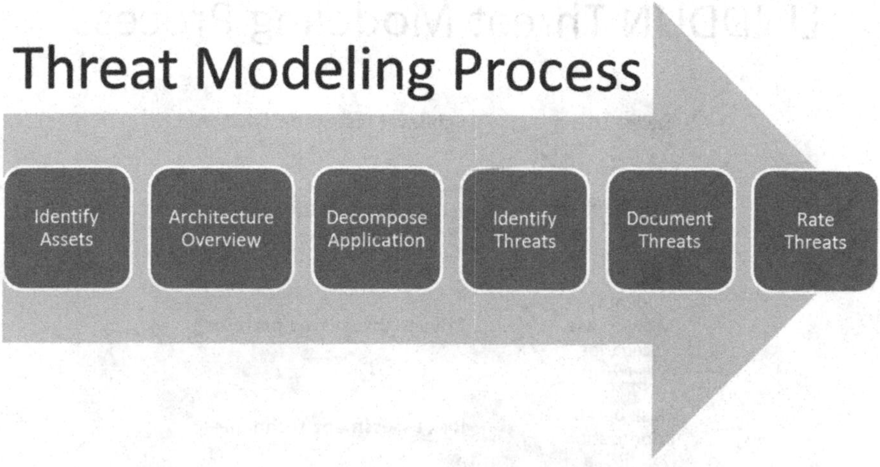

Threat Modeling Process

Fig. 2 Microsoft's threat modeling process. (Adapted from OASIS 2013)

level of security mechanisms in place, and that personal information is protected from a security perspective. While the focus of this chapter is on privacy, we will assume that the system has a reasonable level of security.

The Open Web Application Security Project (OWASP) has developed their own Application Threat Model (OWASP 2013) which has some similarities to Microsoft's model. Based on this model, OWASP has also developed a Mobile Threat Model (Open Web Application Security Project (OWASP) 2013) to identify security threats specifically for mobile applications. OWASP also recommends Microsoft's STRIDE model for identifying threats. We found this useful to consider in a threat model for toy computing, which occurs in a mobile environment. However, this model once again has little focus on privacy.

While both of the above models are primarily concerned with security, there is no widely adopted framework for modeling privacy threats, especially in mobile environments (Roosa 2012). Deng et al. (2010) have developed a methodology called LINDDUN which provides a comprehensive privacy threat modeling framework. Figure 3 illustrates the threat modeling steps in the LINDDUN methodology.

Like the STRIDE and OWASP models, LINDDUN identifies privacy threats by data flow elements and maps them to privacy threats. Misuse case scenarios and privacy threat tree patterns illustrate privacy attack scenarios, which are then prioritized through risk assessment techniques. In the final two steps of this method, mapping the privacy threats to privacy requirements allows for the identification of privacy enhancing solutions. The LINDDUN methodology provides an excellent framework for modeling privacy, although it was not designed for mobile applications in particular. LINDDUN was adapted from the STRIDE model (Meier et al. 2003), using similar threat modeling principles (data flow diagrams, threat trees and trust boundaries) and mapping them to privacy properties based on the terminology defined by Pfitzmann and Hansen (2010). These privacy threats described below are the basis of the LINDDUN methodology and widely recognized in the privacy research community:

LINDDUN Threat Modeling Process

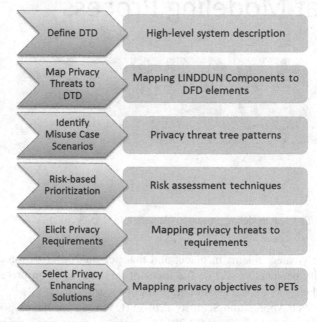

Fig. 3 LINDDUN threat modeling process. (Adapted from Waters et al. 1999)

1. **Linkability**—An attacker is able to distinguish whether two or more items of interest (e.g. subjects, messages, actions, etc.) are related or not within the system.
2. **Identifiability**—An attacker can sufficiently identify a subject associated to an item of interest, for example, the sender of a message. Usually, identifiability refers to a set of potential subjects. This is a special case of linkability between a subject and its attributes. Identifiability is a threat to anonymity and pseudonimity.
3. **Non-repudiation**—This allows an attacker to gather evidence to counter the claims of the repudiating party and to prove that a user knows, has done or has said something.
4. **Detectability**—An attacker can sufficiently distinguish whether an item exists or not (e.g. messages are sufficiently discernible from random noise).
5. **Information Disclosure**—Personal information is exposed to individuals who are not supposed to have access to it.
6. **Content unawareness**— A user is unaware of the information disclosed to the system. The user either provides too much information which allows an attacker to easily retrieve the user's identity or inaccurate information which can cause wrong decisions or actions.
7. **Policy and Consent Noncompliance**—This means that even though the system shows its privacy policies to its users, there is no guarantee that the system actually complies to the advertised policies. Therefore, the user's personal data might still be revealed.

Fig. 4 Threat modeling process

The above threats can be categorized into hard or soft privacy threats (Deng et al. 2010). Our focus for this chapter is on soft privacy: information disclosure and content awareness. Soft privacy is based on the assumption that the data subject is not in control of personal data, and must trust the data controllers (service providers). This is the domain of policies, access control and audit. In this model, the data subject provides personal data and the data controller is responsible for it. Policy consent and noncompliance is beyond the scope of this work, which assumes that the system complies with its privacy policies.

Our Approach

Based on the above threat modeling techniques, we have adapted our own technique appropriate for modeling privacy threats in a mobile toy computing environment. Below is the threat modeling process we will be covering in the following sections, adapted from Microsoft's Threat Modeling Principles (Meier et al. 2003) and STRIDE Model (Hernan et al. 2006), OWASP's Mobile Threat Model (Open Web Application Security Project (OWASP) 2013), and the LINDDUN methodology for privacy threat analysis by Deng et al. (2010). We believe that this will provide an effective analysis of privacy threats in a mobile toy computing environment.

Our approach, illustrated in Fig. 4, uses a similar process as the three models discussed, with greatest motivation from LINDDUN. Starting with an overview of the technical architecture, we will identify location data assets and data flow. Next we will use the LINDDUN methodology to identify privacy threats and threat agents, and illustrate methods of attack through threat trees. Lastly we will use this analysis to identify privacy requirements and controls to mitigate threats to location privacy.

Architecture Overview

A toy computing application allows the user to interact with a physical toy device along with a mobile device to play a game. From an architectural perspective, we will consider the end user components (the user, physical toy component, and mobile device) as one entity, which in our diagram will be referred to simply as the *mobile device*. The user is the individual who is playing with the toy, which is connected to a mobile device also operated by the user. The user interacts with the physical toy and/or mobile device through touch screen, microphone, camera, and/or other sensors such as the accelerometer. The physical toy component may or may not have embedded systems, but must be able to interact in some way with the mobile device (ex. physically, visually, audibly, or through a wireless interface). The toy computing environment follows the BYOD model, where the mobile device is provided by the user and may take the form of a smartphone or tablet. GPS location data is also collected and stored on the mobile device. Data is stored on the device in flash memory and/or removable storage (i.e. SD card), and communicated over wireless interfaces such as Wi-Fi, Bluetooth, NFC, or RFID.

Assets and Data Flow

Identify Assets

With a focus on location privacy, potential sensitive data that could be collected on the user is as follows. Location data is collected through the GPS on the mobile device. As discussed in Chapter "Toy Computing Background", location can be expressed as absolute, relative, or type of location. For the purpose of this scenario, we are concerned with absolute location, which is the location expressed in a range or exact GPS coordinates, latitude and longitude. The absolute location can be expressed as coarse or fine (Android 2015). The location can be collected as a single GPS location event (the location of the user at one point in time), or a GPS trace (a series of location events recorded over a period of time, showing location history). A GPS location includes a timestamp for the time it was detected. The location of the user may be directly or indirectly associated with their real identity and other profile data. In the case of toy computing, the user is a child under the age of 13, their personal information is particularly sensitive, especially when associated with their real identity. Alternatively, location data may be anonymous or associated with a pseudonym. This depends on the architecture and privacy practices of the specific application and service provider, and is beyond the scope of this chapter. Our focus is on enabling the user to be in control of their privacy by specifying their privacy preferences, under the assumption that the service has published an accurate privacy policy and also complies with it. Inferences can be made based on GPS location combined with other data such as:

- Type of motion (walking, standing, running, driving, etc.)
- Interactions with other users, friends in same location (e.g. who they are with at a certain time/day)
- User's real identity associated with their location (name, age, profile data)
- Behavior (e.g. religious beliefs based on going to church, or health based on frequent doctor's appointments)
- Location of home, school or daycare
- Travel patterns (e.g. where they are likely to be at a given time or day)
- If a child is home and parents may not be (e.g. during work hours), or alternatively if a child is in a public place while a parent may not be

From the service provider's perspective, it is necessary to collect this data for the purpose of running the game. Parents who wish to be in control of their child's private location data may limit the collection of this data under certain contexts depending on a company's privacy policies.

Data Flow

Use Case Application—Blaster Toy First-Person-Shooter (FPS) Game

When a game starts, the mobile device connects to mobile service(s) which support gameplay. For the case of this study with a focus on location privacy, the mobile service receives GPS location data on the user and responds with location-based services such as the location of other players. Also connected to the service are other players, and other potential third parties. A blaster toy first-person-shooter (FPS) game, such as *Tek Recon* (discussed in Chapter "Toy Computing Background"), will be used as a use case application throughout this section. *Tek Recon* a multi-player mobile toy computing environment, where users participate in a game with their friends through a mobile application over a network. With our focus on location privacy, the game uses GPS location data from all of the users' mobile devices so they can locate each other on the map.

In the DFD illustrated in Fig. 5, the user is represented as a 3-tuple entity (user, toy, mobile device) which interacts with the system. The toy computing environment contains two processes: the mobile service and the walled garden module. While a user is engaged in the game, they interact with the physical toy and the mobile device, and the mobile device is connected to a mobile service, which may be

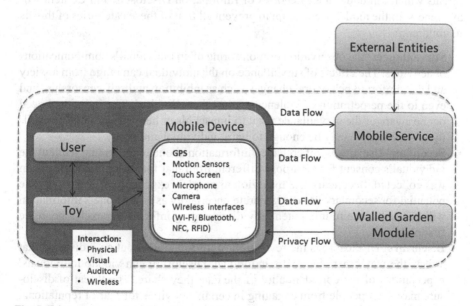

Fig. 5 Tek recon game data flow diagram privacy threats

connected to other entities over a network. GPS location data is collected from the user from the GPS on the device. When the mobile service sends a request for the location data, the mobile device forwards the request to the walled garden module, which checks the policies and makes an access control decision for the request. If the request is permitted, the mobile device then responds to the mobile service with the requested location data.

Privacy Threats

Identify Privacy Threats

From a policy perspective, any data sharing practices that may result in any of the above LINDDUN threats should be identified in the system's privacy policy. This work depends heavily on the assumption that the service has published an accurate privacy policy and also complies with it. For the purpose of this chapter, we aim to address the threats of information disclosure and content unawareness. Information disclosure occurs when a user's personal information is exposed to individuals who are not supposed to have access to it. For the purpose of this work we will assume that although the information disclosure practices are outlined in the privacy policy, and the user has provided their consent, the user is not actually aware due to the face that they did not read or understand the policy. Content Unawareness occurs when the user is unaware of the information that is collected on them, for example their location information. Looking at these threats in more detail, the IETF's RFC6973 on Privacy Considerations (Cooper et al. 2013) provides more specific secondary threats which fall under the categories of Information Disclosure and Content Unawareness. In the model, we attempt to prevent all four of these categories of threats to children:

- **Surveillance**: "the observation or monitoring of an individual's communications or activities. The effects of surveillance on the individual can range from anxiety and discomfort to behavioral changes such as inhibition and self-censorship, and even to the perpetration of violence against the individual. The individual need not be aware of the surveillance for it to impact his or her privacy—the possibility of surveillance may be enough to harm individual autonomy."
- **Secondary Use**: the use of collected information about an individual without the individual's consent for a purpose different from that for which the information was collected. Secondary use may violate people's expectations or desires. The potential for secondary use can generate uncertainty as to how one's information will be used in the future, potentially discouraging information exchange in the first place.
- **Disclosure**: Disclosure is the revelation of information about an individual that affects the way others judge the individual. Disclosure can violate individuals' expectations of the confidentiality of the data they share. The threat of disclosure may deter people from engaging in certain activities for fear of reputational harm, or simply because they do not wish to be observed.

Table 3 Mapping blaster FPS game DFD elements to privacy threats

Entity	User
Process	Game
	Service
Data Store	User's location resources on mobile device
	Service database (DB)
Data Flow	User data stream (user to game)
	Service data stream (game to service)
	DB data stream (service to DB)

- **Exclusion:** Exclusion is the failure to allow individuals to know about the data that others have about them and to participate in its handling and use. Exclusion reduces accountability on the part of entities that maintain information about people and creates a sense of vulnerability in relation to individuals' ability to control how information about them is collected and used.

Mapping Privacy Threats to DFD

Referencing the DFD from Fig. 5, we will now outline the DFD elements and then map the privacy threats to the DFD. Table 3 shows the DFD elements in the Blaster FPS Game mentioned in the previous section.

Now based on the above DFD elements, in Table 4 we map the LINDDUN privacy threats to DFD element types (E: Entity, DF: data flow, DS: data store, P: process) in a toy computing scenario with the Tek Recon example:

The threat of information disclosure occurs at the process, data store, and data flow levels. This falls into the control of the service provider, who outlines information disclosure practices in their privacy policy. While we assume that the service has accurate policies and also complies with them, the threat we are concerned with is then with the entity who agrees to disclose the information. Content unawareness

Table 4 Mapping privacy threats to DFD elements

Threat Categories	Entity	Process	Data store	Data flow
Linkability	X	X	X	X
Identifiability	X	X	X	X
Non-repudiation		X	X	X
Detectability		X	X	X
Information disclosure	T	A	A	A
Content unawareness	T			
Policy/consent noncompliance		A	A	A

Legend: *x* Out of scope, *T* Threats addressed, *A* Assumed to comply is a threat to the entity (user). The user is required to provide the necessary consent to process personal data. The goal of our model is to address the threats of Content Unawareness from the perspective of the user, putting them in control of information disclosure. This model will address information disclosure from the entity (user)'s perspective who complies with information disclosure practices. This model is acting under the assumption that the process, data store, and data flow elements all act in compliance with their policies and the consent of the user.

Methods of Attack

In this section we will observe different methods an attacker can use to reach the data. First we will examine privacy threats based on Table 3 in the previous section to determine privacy threat trees. Next, we will create misuse case scenarios based on the threat tree patterns.

Privacy Threat Trees

Information Disclosure

Figure 6 refers to the privacy threat tree for information disclosure. For the purpose of this work, we are referring to intentional information disclosure, which is predefined by the service and outlined in the privacy policy, rather than information disclosure as a result of security exploits. Information disclosure can occur at process, data store, or data flow level. Location information may be disclosed to other users or with a third party. The threats related to sharing an entity's location data can lead to undesirable inferences of the user's behavior and personal life (see list of inferences in Section "Assets and Data Flow"). A child's location data sent to

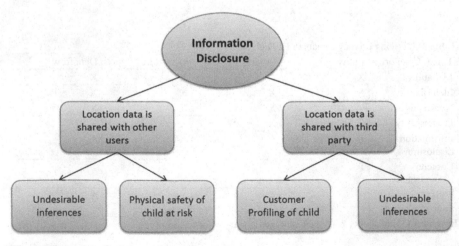

Fig. 6 Information disclosure privacy threat tree

a third party can be used for customer profiling of the child. Sharing location data with other users puts the physical safety of the child user at risk if it is shared with an untrusted entity. For these reasons, a user may choose not to consent to sharing their location data depending on privacy policy practices.

Content Unawareness of Entity

Content unawareness occurs at the user level when the user provides more personal data than is required, or does not read the privacy policies. Providing too much personal data is unnecessary and opens up opportunity for further undesirable inferences. It is also possible that a user does not read the privacy policies and therefore is unaware that certain aspects of their personal data is being collected and shared. The user may be unaware of the purpose their location data is collected, or how it is used. The user may not even be aware that their location information is being collected at all. Additionally, the user may not be aware that their location data is being shared with third parties. All of these situations can result in information disclosure (see previous section) to which the user has unknowingly provided their consent (Fig. 7).

Misuse Case Scenarios

In this section we provide a misuse case scenario of Tek Recon based on the threat tree patterns in the previous section. The misuse case model is based on the LIND-DUN model.

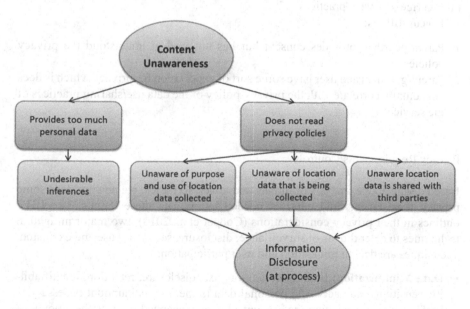

Fig. 7 Content unawareness privacy threat tree

Content Unawareness and Information Disclosure

The threat trees in the previous section indicates that in order to be susceptible to the threat of content awareness, the user either unknowingly provides too much personal data, or does not read privacy policies. For information disclosure, the mobile service forwards the data to a third party or another user. These are the preconditions of the misuse case. To create the attack scenarios, the attacker first needs to have access to the data store, and either the user (data subject) can be re-identified or the pseudonyms can be linkable. In this scenario, the actions of the misactor are actually completely legitimate as outlined in their privacy policy, however the data use/sharing practices do not comply with the user's expectations or legislation.

Title: Misuse Case 1: Content Unawareness and Information Disclosure

Summary: User unknowingly provides location data to the service

Assets, stakeholders and threats: location information of the user. The parent/guardian and user are unaware the information is collected and sent. Potential threats: surveillance, secondary use, disclosure, exclusion

Primary Misactor: Parent/guardian for not reading privacy policy.

Basic Flow:

1. Parent/guardian consents to privacy policy without reading it.
2. User unknowingly sends location information to the mobile service.

Alternative Flow:

1. Same as the above, and the mobile service sends user's location information to a third party for marketing purposes.

Trigger: Parent/guardian does not read the privacy policy which outlines the mobile service's privacy practices.

Preconditions:

1. Parent/guardian provides consent but has not read or understood the privacy policies.
2. Parent/guardian and user have some sort of expectation for privacy which is does not actually correlate with the privacy policy or the data use/sharing practices of the service.

Privacy Requirements/Controls

Based on the above analysis of threats and attack scenarios, we now propose some privacy requirements and controls needed to mitigate these threats. The IETF outlines in their privacy considerations (Cooper et al. 2013), two major mitigation techniques to deter threats of surveillance, disclosure, secondary use and exclusion. Techniques are data minimization and user participation:

- **Data Minimization:** limiting collection, use, disclosure, retention, identifiability, sensitivity, and access to personal data to the minimal amount necessary to perform a task. Reducing the amount of data exchanged reduces the amount of data that can be misused or leaked. Data Minimization mitigates the threats of: surveillance, secondary use, and disclosure.

- **User Participation**: data collection and use that happens "in secret," without the individual's knowledge, is apt to violate the individual's expectation of privacy and may create incentives for misuse of data. As a result, privacy regimes tend to include provisions to support informing individuals about data collection and use and involving them in decisions about the treatment of their data. In an engineering context, supporting the goal of user participation usually means providing ways for users to control the data that is shared about them. It may also mean providing ways for users to signal how they expect their data to be used and shared. User participation mitigates the threats of: surveillance, secondary use, disclosure, and exclusion.

Our threat model illustrates that the privacy requirements for toy computing are data minimization and user participation, in order to mitigate the threats of information disclosure and content unawareness, which can lead to surveillance, disclosure, secondary use and exclusion. Privacy controls which achieve the goals of data minimization and user participation include implementing a privacy access control model.

Section 3. Privacy Considerations

End User Requirements for Children

Children provide a unique user base which requires special attention in several key areas related to their privacy. Firstly, it is widely accepted internationally that that a child's data is considered particularly sensitive and should be treated with extreme care (Office of the Privacy Commissioner of Canada 2014) (International Telecommunication Union (ITU), United Nations Children's Fund (UNICEF) 2014). Online privacy for children has been a great concern, and this concern is inherited into the toy computing environment, particularly when the child's location is involved and can be potentially shared with other parties. Children must be protected from violence, sexual abuse and exploitation which they can be vulnerable to online including harassment, stalking, grooming, sexual abuse or exploitation, or personal data misuse (Livingstone et al. 2011). Sexual solicitation and internet-initiated offline encounters are a major issue for the online safety of children (Schrock and Boyd n.d.). The U.S. Department of Justice (U.S. Department of Justice, National Sex Offender Public Website n.d.) indicates that "1 in 25 youths received an online sexual solicitation in which the solicitor tried to make offline contact". All of these risks are increased with the possibility of a potential solicitor becoming aware of the child's location or historical location patterns. On the other hand, children also take up a large segment of the consumer population and are of particular interest to market researchers who may attempt to collect their personal data and usage patterns for targeted advertising (Salomon 2010). Third party advertisers can infer a great amount of information about a child based on their location and other context information, collecting detailed behavioral profiles that may be used for unknown or unwanted purposes.

Another concern with child users is that the usage patterns of children differ from that of an adult. Children often have little understanding or regard for the privacy of their information, and are more likely to act in spontaneous ways. The usage behavior of children indicates that they are more open to giving out personal information, which makes issues of sensitive data sharing of great concern. Child users exhibit a varying level of awareness when it comes to their online activities and understanding of privacy risks. There is a fluctuating level of comfort and knowledge with technology and online activities among children, where usage and online behavior differ according to their age, development level, and frequency of use. Children may lack the maturity to appreciate the wider social and personal consequences of revealing or agreeing to share their personal information online, or the use of their personal information for commercial purposes (International Telecommunication Union (ITU), United Nations Children's Fund (UNICEF) 2014). Younger children in particular generally lack the skills and confidence in areas of internet use that are especially important for safety (Livingstone et al. 2011). In order to be effective, privacy protection strategies are required which incorporate measures and messages appropriate to different ages and levels of understanding (United Nations Children's Fund (UNICEF) 2011).

The online mobile environment and associated toy computing technology provides many opportunities for children, but is also accompanied by several risks. Many initiatives have been undertaken in attempt to provide an online environment which is safe and age-appropriate, and to help children to be empowered and engaged in the online environment (European NGO Alliance for Child Safety Online (eNASCO) 2010). Organizations such as UNICEF have been working to promote digital citizenship among children and develop products and platforms that facilitate children's positive use of technology (International Telecommunication Union (ITU), United Nations Children's Fund (UNICEF) 2014). Noted by Westin (1967), "each individual is continually engaged in a personal adjustment process in which he balances the desire for privacy with the desire of disclosure and communication of himself to others." Several works such as (Chakraborty et al. 2013) iterate the competing goals of utility vs. privacy. A popular theme in industry guidelines indicates a need for balance between children's right to protection from violence, sexual abuse and exploitation, and the right to information access, freedom of expression, privacy and non-discrimination (United Nations Children's Fund (UNICEF) 2011). Measures for protection must also not be overly restrictive for the child or other users (International Telecommunication Union (ITU), United Nations Children's Fund (UNICEF) 2014). It is possible for some measures to challenge current business models, reduce the competitiveness of a company, or threaten other freedoms currently observed online (United Nations Children's Fund (UNICEF) 2013). The Toy Industry Association (TIA) has raised concerns that restrictions could limit the ability for toy companies to obtain necessary data to analyze and improve content, allow children to enjoy personalized but anonymous online experiences, and benefit from the ability to offer targeted advertising on their e-commerce and adult sites (Toy Industry Association 2012). Thus it is necessary to find the appropriate balance between privacy and freedoms to users.

Parental Control

A report from Pew Research Center and Berkman Center for Internet & Society at Harvard University (Madden et al. 2012) indicates that the majority of parents in the United States are concerned about their children's online privacy, some of the main concerns being related to strangers online, and the data advertisers are collecting about their children's online behavior. While parents strive to ensure their child's physical and online safety and privacy, they may wish to be in control of how their personal data is shared through the devices they are using. The main protection children have on the Internet is parental guidance and supervision (Salomon 2010). Privacy controls can allow parents to create policies to prevent their children from allowing their data to be collected from services according to their preferences. The ITU and UNICEF recommend parental controls "not to transfer responsibility for children's ICT use to parents alone, but to recognize that parents are in a better position to decide what is appropriate for their children and should be aware of all risks in order to better protect their children and empower them to take action" (International Telecommunication Union (ITU), United Nations Children's Fund (UNICEF) 2014). Parental controls can be rated by their functionality, effectiveness, usability, and security (European Union Safer Internet Program n.d.):

- **Functionality:** Does the tool have the required functions for the parent's needs?
- **Effectiveness:** Does the tool successfully block the intended content or action?
- **Usability:** Is the tool easy to install, configure and use?
- **Security:** Does the tool prevent the child from bypassing or disabling the controls?

In regards to usability, parents sometimes have less understanding and knowledge of the internet and mobile devices than their children. Further, convergence of mobile devices and internet services makes parental oversight more difficult (International Telecommunication Union (ITU), United Nations Children's Fund (UNICEF) 2014). This introduces difficulties with a parent's ability to effectively implement privacy controls for their child. In a mobile toy computing environment, the child will likely be using either a mobile device belonging to his/her parent or using his/her own mobile device. Parents may face privacy challenges related to reviewing privacy policies of applications that they or their children may be using. A study by Chin et al. (2012) of 60 participants found that users are highly likely to install free applications, and place a higher value on other user reviews than of privacy policies and EULAs. Also, users often have more dangerous tendencies on their mobile behavior than they do on a computer or laptop. When reviewing privacy policies, parents are likely to face difficulty in reading and understanding the policies. A further study by Felt et al. (2012) found that only 17 % of participants paid attention to permissions during installation, and only 3 % could correctly answer permission comprehension questions. It can be inferred from this research that the majority of users do not understand or care about the permission warnings that they receive on their mobile devices. This is a huge disadvantage to the current permissions system, illustrates the need to improve the usability of privacy protecting systems. Peng et al. (2012) discuss the importance of communicating the privacy risks of an application to users, while also proposing a method for ranking risks based on probabilistic generative models. It is understandable that parents will run into similar issues with understanding privacy practices in regards to their children. While they will also likely be even more concerned with their child's privacy, it is important to parents/guardians that they are able to understand and correctly control their child's private data. A privacy preserving framework is required to allow parents to easily and effectively set preferences to control and restrict the personal data that can be collected on their child.

Privacy Laws and Regulations

Privacy protection laws define the rights of data subjects (users), the responsibilities of data collectors (service providers), and methods for dispute resolution. These laws are generally enforced through ombudsmen (e.g. Privacy Commissioner of Canada), or licensing bureaus (e.g. CNIL in France) (Reay et al. 2009). Different countries and legislations have different laws for privacy protection, and there are also many international guidelines and industry regulations which outline privacy best practices. These laws and regulations can also differ depending on what type of information is being collected (e.g. health information), or who the users are (e.g. children under the age of 13). In this chapter, we aim to investigate the privacy aspects relevant to the child users, toys and safety, and location data.

Toy computing encompasses a range of technologies, including traditional and electronic toys, internet, mobile devices, and inherits the privacy requirements and governing laws and regulations of each, with particular attention to children. Traditional distinctions between different parts of the telecommunications and mobile phone industries, and between internet companies and broadcasters, are fast breaking down or becoming irrelevant (International Telecommunication Union (ITU), United Nations Children's Fund (UNICEF) 2014). With the change in technology, regulating bodies have recently been striving to update laws, regulations and industry guidelines. Table 5 shows the privacy concerns associated with each technology and the corresponding laws and regulating organizations to address them.

Privacy Principles

Canada's privacy laws are outlined in The Personal Information Protection and Electronic Documentation Act (PIPEDA) (Canadian Public Works and Government

Table 5 Privacy concerns and regulation across toy computing components

Component	Traditional & electronic toys	Internet/web services	Mobile apps/services
Children's privacy concerns	Physical safety	Inappropriate content, conduct, or contact	Pervasiveness
			Location
			BYOD
Laws and regulation	Toy safety guidelines	PIPEDA (Canada)	MMA
	Toy Industry Association (TIA)	COPPA (USA)	CTIA
		UNICEF/ITU Industry guidelines on child online protection	

Services 2000), which governs how personal information can be collected, used, and disclosed in commercial business. PIPEDA is based on the 10 principles of privacy outlined in the Canadian Standards Association's (CSA) Model Code for the Protection of Personal Information (Government of Canada 2000), which has been recognized as a national standard as of 1996 (Canadian Standards Association 1996). This model code is representative of principles behind privacy legislation in many countries, including the United States and the European Union. It also bears similarities to the Organization for Economic Cooperation and Development (OECD) Guidelines for the Protection of Privacy and Transborder Flows of Personal Data (Organization for Economic Cooperation and Development 2013) which have been adopted by member countries of the European Union (European Parliament and of the Council of the European Union 1995). The CSA's 10 Principles of Privacy are summarized as follows (Canadian Standards Association 1996):

1. **Accountability**—an organization is responsible for personal information under its control and shall designate an individual or individuals who are accountable for the organization's compliance with the following principles.
2. **Identifying Purposes**—the purpose for which personal information is collected shall be identified by the organization at or before the time the information is collected.
3. **Consent**—The knowledge and consent of the individual are required for the collection, use, or disclosure of personal information, except when appropriate.
4. **Limiting Collection**—the collection of personal information shall be limited to that which is necessary for the purposes identified by the organization. Information shall be collected by fair and lawful means.
5. **Limiting Use, Disclosure, and Retention**—personal information shall not be used or disclosed for purposes other than those for which it was collected, except with the consent of the individual or as required by the law. Personal information shall be retained only as long as necessary for fulfillment of those purposes.
6. **Accuracy**—personal information shall be as accurate, complete, and up-to-date as is necessary for the purposes for which it is to be used.
7. **Safeguards**—personal information shall be protected by security safeguards appropriate to the sensitivity of the information.

8. **Openness**—an organization shall make readily available to individuals specific information about its policies and practices relating to the management of personal information.
9. **Individual Access**—upon request, an individual shall be informed of the existence, use and disclosure of his/her personal information and shall be given access to that information. An individual shall be able to challenge the accuracy and completeness of the information and have it amended as appropriate.
10. **Challenging Compliance**—an individual shall be able to address a challenge concerning compliance with the above principles to the designated individual or individuals for the organization's compliance.

Children's Privacy Laws

PIPEDA does not mention any special regulations for children. While PIPEDA requires meaningful consent for the collection of personal data collection, it does not refer to a particular age threshold for this. There is a difficulty in determining if a child is able to provide meaningful consent, as this greatly depends on their cognitive and emotional development and their understanding of privacy and online practices (Office of the Privacy Commissioner of Canada 2014). The Office of the Privacy Commissioner of Canada (OPC) has indicated in the Online Behavioural Advertising Guidelines (Office of the Privacy Commissioner of Canada 2012), a focus towards protecting children's online privacy particularly in the region of online behavioural targeted advertisements. The OPC recommends for organizations to avoid knowingly tracking children and websites aimed at children. The OPC has also made the following recommendations regarding the management of the personal information of children and youth (Office of the Privacy Commissioner of Canada 2014), however these recommendations are not legally binding:

- Children's information is considered sensitive and merits special consideration under privacy laws.
- Organizations should implement innovative ways of presenting privacy information to children and youth that take into account their cognitive and emotional development and life experience.

The United States Federal Trade Commission (FTC) Children's Online Privacy Protection Act (COPPA) (United States Federal Trade Commission 1998) protects the online privacy of children under the age of 13, and indicates that a child's personal information cannot be collected without parental consent. In 2010, an amendment to COPPA further elaborated that personal information includes geolocation information, photographs, and videos. In the European Union, privacy laws are governed by the European Union Data Protection Directive (EUDPD) 95/46/EC (European Parliament and of the Council of the European Union 1995), which also has special considerations for children under the age of 13. The Directive states that consent must be given by the child's parent or custodian, and must also be verifiable (Article 8). Further, the United Nations Convention on the Rights of the Child (CRC) (United Nations Children's Fund (UNICEF) 1989) is the most widely endorsed international human rights treaty. This treaty protects children from all forms of violence, exploitation and abuse and discrimination, and ensures that the child's best interest should be the primary consideration in any matters affecting them. UNICEF defines children as individuals under the age of 18, according to Article 1 of the Convention on the Rights of a Child (International Telecommunication Union (ITU), United Nations Children's Fund (UNICEF) 2014). Canada does not have an equivalent to COPPA, or any specific mention of children within PIPEDA. For the purpose of this work we will consider the recommendations by the OPC to preserve children's privacy. We will follow the direction of COPPA and the EUDPD in the definition of a child as an individual under the age of 13 years old. With this in mind, we aim to assist the parent/guardian in protecting the privacy of their child by putting them in control of the information of their child that is shared.

Location Privacy Laws

There do not appear to be special laws in place for regulating the privacy aspects of location data, however it is categorized as personal information is therefore covered by PIPEDA. PIPEDA recognizes personal information in Section 2(1) as "information about an identifiable individual, but does not include the name, title or business address or telephone number of an employee of an organization." PIPEDA defines a "record" as including "any correspondence, memorandum, book, plan, map, drawing, diagram, pictorial or graphic work, photograph, film, microform, sound recording, videotape, machine-readable record and any other documentary material, regardless of physical form or characteristics, and any copy of any of those things". An "electronic document" is defined by PIPEDA is "data that is recorded or stored on any medium in or by a computer system or other similar device and that can be read or perceived by a person or a computer system or other similar device. It includes a display, printout or other output of that data." GPS data is not explicitly mentioned in either of these definitions, however it is possible that it could be considered included. In a 2006 PIPEDA case summary, tracking information collected from a GPS placed in company vehicles was acknowledged as personal information, "since the information can be linked to specific employees driving the vehicles. The employees are identifiable even if they are not identified at all times to all users of the system" (PIPEDA Case Summary #2006-351). In 2010, an amendment to COPPA further elaborated that personal information includes geolocation information, photographs, and videos. For the purpose of this work, we will consider COPPA's reference to geolocation information as a type of personal information to be protected.

Health Canada's Safety Requirements for Children's Toys and Related Products

In Canada, the Canada Consumer Product Safety Act (CCPSA) applies to all manufactures, importers, advertisers, sellers, or testers of a consumer product. A toy is defined by the Government of Canada's Toy Regulations (Health Canada 2011) as "a product that is intended for use by a child in learning or play". Health Canada identifies a toy as intended for use by children less than 14 years of age, unless a younger age is prescribed within a requirement. This is consistent with several international toy safety standards (Health Canada 2012):

* ISO 8124-1 Safety of toys—Part 1: Safety aspects related to mechanical and physical properties
* EN71-1 Safety of toys—Part 1: Mechanical and physical properties
* ASTM F963 Standard Consumer Safety Specification for Toy Safety

These safety requirements were designed for traditional toys and outline primarily physical safety hazards (e.g. mechanical hazards, electrical hazards, etc.). Electric toys have specific safety standards as well, and must meet the requirements as set out in Canadian Standards Association Standard C22.2 No. 149-1972, *Electrically Operated Toys*, (Sect. 5 of the *Toys Regulations* 2011). While safety concerns concentrate primarily on physical safety limited to the physical design of the toy, privacy is not a topic widely addressed in the toy industry. Safety is a concern related to privacy when a breach of privacy can result in physical harm to the child. Physical threats to safety as a result of privacy breach can include exploitation, stalking, physical harm that can happen as a result of the knowledge of a child's location. Current toy safety regulations are not up to date and do not acknowledge the possibility of this type of safety threat.

Industry Guidelines and Best Practices

While technology continues to change, there are limitations on privacy laws and many countries and states struggle to keep up with the changing environment. Privacy related to location information, child users, and responsible marketing to children have been emerging topics in recent years. In order to help regulate this, several regulating organizations have provided guidelines and recommendations for industry *self-regulation* of the management of children's data online and mobile environments. The International Telecommunication Union (ITU) and United Nations Children's Fund (UNICEF) have released guidelines for child online protection, stressing that companies in states which lack adequate legal frameworks for the protection of children's rights to privacy and freedom of expression should follow enhanced due diligence to ensure policies and practices are in line with international law (International Telecommunication Union (ITU), United Nations Children's Fund (UNICEF) 2014). The guidelines encourage companies to adopt the highest privacy standards when it comes to collecting, processing and storing data from or about children (International Telecommunication Union (ITU), United Nations Children's Fund (UNICEF) 2014). Further, services directed at or likely to attract a main audience of children must consider the risks posed to them by access to, or collection and use of, personal information (including location information), and ensure those risks are properly addressed (International Telecommunication Union (ITU), United Nations Children's Fund (UNICEF) 2014).

The Mobile Marketing Association (MMA) has issued a Mobile Application Privacy Policy Framework (MMA Privacy & Advocacy Committee 2011) to help mobile application providers to create privacy policies. This document has special considerations for location information and children. If a child has provided information without their parent's consent, the parent can contact the provider to delete the information. The CTIA recommends Best practices and Guidelines for location based services (CTIA—The Wireless Association 2010) which follow closely with the privacy principles. This document does not appear to have much entirely unique for location information, but there are a couple notable items. The CTIA outlines that user's location information should be retained by LBS providers only as long as required by business needs, after which time it must be destroyed. If location must be retained for long-term use it should be converted to aggregate or anonymized data. The importance of the protection of minors is also elaborated by the CTIA regarding the use and disclosure of location information. The Digital Advertising Alliance (DAA) has also issued a report on the Application of Self-Regulatory Principles to the Mobile Environment (Digital Advertising Alliance 2013). The focus of this report is on transparency and control, and includes special considerations for "Precise Location Data".

The North American Toy Industry Association (TIA) released a whitepaper (Toy Industry Association 2012) regarding the changing privacy and data security landscape the toy industry is facing with the emerging popularity of child-directed mobile apps. The TIA iterates the issues of children's marketing and privacy in this context, indicating that privacy and data security issues affect day to day operations of toy companies. The TIA offers the following concerns:

- The FTC's restrictions on third party sharing, except where information is used to support the internal functions of the website, could restrict routine use of web analytics and other activities currently permitted under the existing rule. The proposed rule affirms that COPPA applies to all online services directed to children, including mobile apps.
- Toy companies support parental authority and strive to offer children interactive, anonymous experiences under the current COPPA framework.
- Self-regulation provides an effective means of protecting children's privacy; intrusive oversight will undercut the effectiveness

Section 4. Privacy Requirements for Toy Computing

In this analysis for privacy in a toy computing environment, there were several factors to take into consideration regarding the privacy goals based on our threat model, end user requirements, laws and regulations. In this section we present the requirements for a privacy framework based on these factors. Data minimization and user participation are privacy goals based on our threat model. A framework is required which can achieve these privacy goals by minimizing the collection and retention of potentially sensitive user data, as well as involving the user (or parent) in the control of their child's data. End user requirements need to consider that the main user base is children, who have unique requirements as they are especially vulnerable and in order to protect their sensitive location data, parents/guardians require a method to implement privacy controls on their child's data. Next, the framework must help to achieve the 10 principles of privacy and comply with PIPEDA.

Six Privacy Constraints for Toy Computing

Based on the above, we have compiled 6 privacy rights for parents/guardians to have control over their child's location data in toy computing. These privacy requirements enforce the goals of data minimization and user participation, by allowing parents/guardians to be in control of how their child's privacy is managed, and restrict the data that is collected. These requirements comply with the 10 principles of privacy of which PIPEDA is based.

1. **The right for a parent/guardian to request restrictions on the use or disclosure of private information of their child**. This allows parents/guardians to provide restrictions to purpose, recipients, obligations, and retention regarding their child's location information. This protects children from having their location information being used or shared for any purpose considered illegitimate or unacceptable to the parent/guardian.

   - *Goals*: user participation, data minimization
   - *Privacy Principles*: consent, limiting collection, limiting use disclosure and retention

2. **The right for a parent/guardian to access, copy, and inspect collected records on their child**. This allows a parent/guardian to access their child's location records to see that data that is collected on them.

   - *Goals*: user participation
   - Privacy Principles: access

3. The right for a parent/guardian to request deletion of their child's private data records, or correction if records are inaccurate. This allows parents/guardians to request that their child's location records be deleted, or to request a correction if their child's location records are incomplete or incorrect.

   - *Goals*: user participation, data restriction
   - *Privacy Principles*: limiting collection & retention, accuracy, access

4. The right for a parent/guardian to request acknowledgements through a communication channel when private information of their child is collected. This allows parents to set up a communication channel such as phone number or email address to receive acknowledgements there is an update pertaining to the

collection of their child's location records. This allows parents/guardians to keep track of how their child's location information.

- *Goals*: user participation
- *Privacy Principles*: openness, access

5. **The right to file complaints to toy company.** If a parent/guardian believes that their child's data has been mishandled in any way by the toy company or service provider, or if they believe that they have not acted in compliance with their policies, they are able to file complaints.

- *Goals*: user participation
- *Privacy Principles*: accountability, challenging compliance

6. **The right to find out where the child's private data has been shared for purposes other than a game.** This allows a parent/guardian to be notified if their child's location records have been shared with another party for any purpose other than for a game.

- *Requirements*: user participation
- *Privacy Principles*: notice, purpose, openness

Section 5. Demo Privacy Preference Interfaces

The earlier sections of this chapter provided an analysis on the various privacy requirements for toy computing. In this section, we present a demo of an interface for parents/guardians to use in an initial setup to configure preferences and create policy rules. These options would appear during initial setup of a toy computing application. These privacy settings allow parents to create access control rules based on their preferences on concepts from P3P (Cranor et al. 2002), i.e. purposes, recipients, obligations, and retentions.

Profile Setup

The first step in the configuration process is the Profile Setup phase. The Profile Setup phase includes three sections, the Parent/Guardian Contact Details, Child Information, and Privacy Policy Review.

Figure 8 shows the first two screens of the Profile Setup Phase. In the first screen, the parent/guardian enters their basic information including name and email address, and then selects if they wish to receive email updates on their child's privacy-related information. Next, on the Child Information page, the child's first name

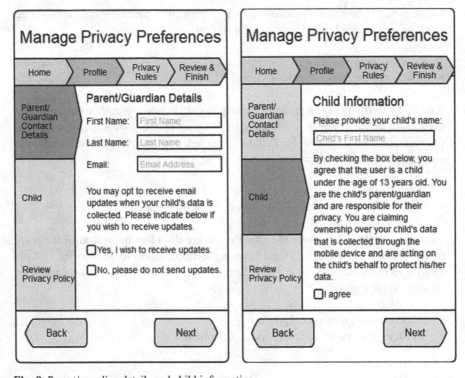

Fig. 8 Parent/guardian details and child information

is entered for management purposes, and the parent/guardian then agrees to take ownership over their child's data.

Next, the privacy policy of the mobile toy application is presented to the parent/guardian to review. The parent/guardian reviews the policy and must confirm that they have read and agree to the terms before proceeding.

Figure 9 shows this screen with the *Tek Recon* privacy policy. By agreeing to the terms, the parent/guardian is providing consent on their child's behalf.

Configuring Privacy Rules

The next phase of the setup is the Privacy Rule creation phase. In this phase, the parent/guardian is able to create one or more privacy rules for how their child's private location data is used. By default there are no policy rules yet configured. As shown in Fig. 10, a new rule can be created or a template can be used. Templates of useful policy rules can be provided to simplify the rule configuration process for parents/guardians. However, in this example we will show how to create an entirely new rule from scratch.

Fig. 9 Review privacy policy

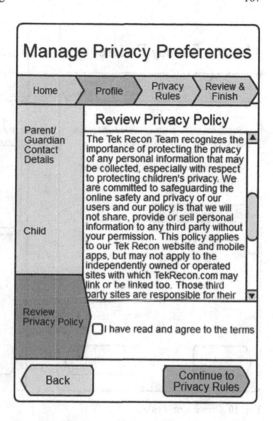

The first step of creating a new privacy rule is the General Settings, as shown in Fig. 11. In the General Settings, the parent/guardian can name the rule, provide a description, and set an expiry date for how long the rule will be in effect. Next, in the Core Access Control settings, the mobile service (subject), location resource (object), and operation are selected. The objects selected are the absolute location and relative location.

Next, the settings for Purposes and Recipients are also presented in Fig. 11. The parent/guardian chooses from a list of purposes they wish to accept, as well as a list of types of recipients. The types of recipients can be expanded to be more specific, such as Third-Party: Marketing, or Group: Game Players.

The next steps are the Obligations and Retention settings, and then finally reviewing and adding the rule, as shown in Fig. 12, the parent/guardian first selects the obligations that the service must comply with upon receiving the child's data. Obligations can include compliance with PIPEDA or COPPA. The parent/guardian can also search from a list of other obligations, or input a custom obligation policy. For retention, the parent/guardian can select how long they wish to allow their child's data to be retained. Finally, on the Review & Add Rule page, the privacy rule is presented in plain English. Once the parent/guardian reviews the rule, they can select "Confirm and Add Rule" at the bottom of the screen.

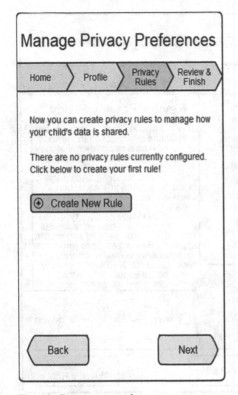 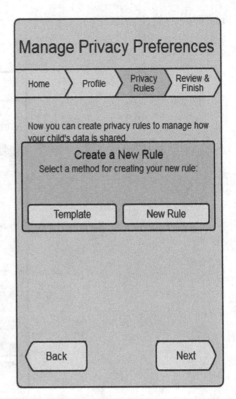

Fig. 10 Create a new rule

Once a privacy rule is added, the parent/guardian is directed to the Manage Privacy Rules page, illustrated in Fig. 13. The Manage Privacy Rules page shows a table of all of the configured privacy rules and their status (e.g. enabled, disabled, or expired). This provides options to enable, disable, edit, delete, or create new rules. A parent/guardian can also return to this screen at a later time to manage rules or renew expired rules. Once the parent/guardian is satisfied with the privacy rules, he/she can select "Next" to be directed to the final Review & Finish page. This page summarizes all of the settings and confirms that the parent/guardian has completed all of the sections. A list of enabled privacy rules and their corresponding expiry dates is also presented. Finally, the parent/guardian can select "Save and Finish" to save their settings and finish the setup. Once the setup is finished, the settings will take effect immediately.

Section 6. Chapter Summary

This chapter outlines the privacy requirements and concerns that parents and children observe in a mobile toy computing environment. We provide an overview of privacy issues, and next present a privacy threat model illustrating the privacy

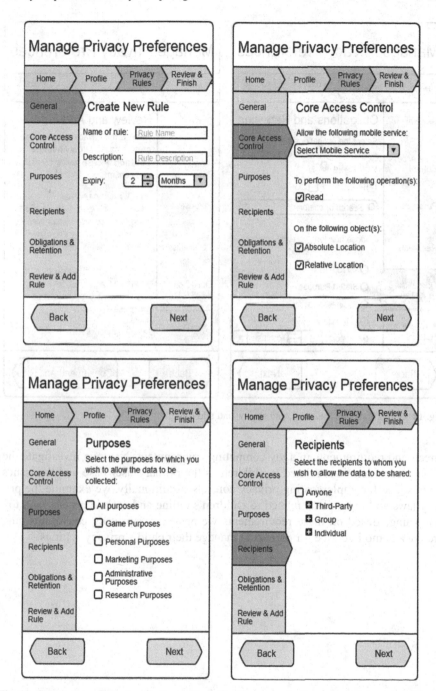

Fig. 11 Create new rule: general, core access control, purposes and recipients

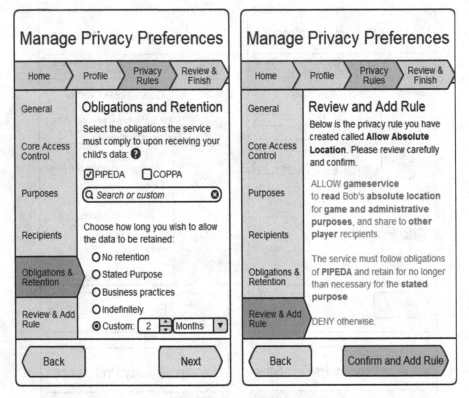

Fig. 12 Obligations and retention, and review and add rule

threats and requirements in a toy computing environment. Next we investigate the privacy considerations related to children as the primary user base and the issues parents face for implementing privacy controls. Additionally, we examine the privacy laws and regulations related to children's online and mobile privacy and toy computing. Based on these requirements we present six privacy constraints, and present a demo interface for parents to manage their child's privacy settings.

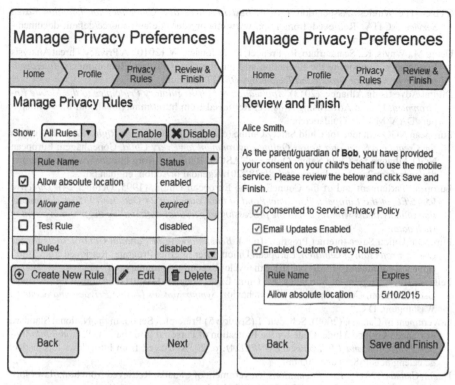

Fig. 13 Manage privacy rules, and review and finish

References

Android. (2015). *Location Strategies*. (Android Developer) Retrieved February 2015, from http://developer.android.com/guide/topics/location/strategies.html

Canadian Public Works and Government Services. (2000). *Personal Information Protection and Electronic Documents Act*.

Canadian Standards Association. (1996). *Archived—Appendix 3: Model Code for the Protection of Personal Information*. Retrieved February 2015, from http://cmcweb.ca/epic/internet/incmc-cmc.nsf/en/fe00076e.html

Chakraborty, S., Raghavan, K., Johnson, M., & Srivastava, M. (2013). A Framework for Context-Aware Privacy of Sensor Data on Mobile Systems. *The Fourteenth Workshop on Mobile Computing Systems and Applications (ACM HotMobile 2013)* (pp. 1–6, Article 11). New York, USA: ACM.

Chin, E., Felt, A., Sekar, V., & Wagner, D. (2012). Measuring User Confidence in Smartphone Security and Privacy. *Symposium on Usable Privacy and Security*. Washington, D.C.

Cooper, A., Tschofenig, H., Aboba, B., Peterson, J., Morris, J., Hansen, M., & Smith, R. (2013). *RFC 6973: Privacy Considerations for Internet Protocols*. IETF. Retrieved from tools.ietf.org/html/rfc6973

Cranor, L., Langheinrich, M., Marchiori, M., Presler-Marshall, M., & Reagle, J. (2002). *The Platform for Privacy Preferences 1.0 Specification*. W3C. Retrieved from http://www.w3.org/TR/P3P/

CTIA—The Wireless Association. (2010). *Best Practices and Guidelines for Location Based Services*. CTIA. Retrieved from http://www.ctia.org/docs/default-source/default-document-library/pdf-version.pdf?sfvrsn=0

Deng, M., Wuyts, K., Scandariato, R., Preneel, B., & Joosen, W. (2010). A Privacy Threat Analysis Framework: Supporting the Elicitation and Fulfillment of Privacy Requirements. *Interdisciplinary Institute for Broadband Technology (IBBT)*. Belgium.

Digital Advertising Alliance. (2013). *Application of Self-Regulatory Principles to the Mobile Environment*. Digital Advertising Alliance. Retrieved from http://mmaglobal.com/files/whitepapers/DAA_Mobile_Guidance.pdf

European NGO Alliance for Child Safety Online (eNASCO). (2010). *The Right Click: An Agenda for Creating a Safer and Fairer Online Environment for Every Child*. Copenhagen: European NGO Alliance for Child Safety Online, eNASCO. Retrieved from http://www.enacso.eu/media/com_form2content/documents/c8/a38/f201/AgendaForAction_english.pdf

European Parliament and of the Council of the European Union. (1995, October 24). *Directive 95/46/EC of the European Parliament and of the Council of 24 October 1995 on the protection of individuals with regard to the processing of personal data and on the free movement of such data*.

European Union Safer Internet Program. (n.d.). *Benchmarking of Parental Control Tools for the Onling Protection of Children*. European Union Safer Internet Program. Retrieved from http://sipbench.eu/transfer/SIP_BENCHII_5th_cycle_Executive_summary.pdf

Felt, A., Ha, E., Egelman, S., Haney, A., Chin, E., & Wagner, D. (2012). Android Permissions: User Attention, Comprehension, and Behavior. *Symposium on Usable Privacy and Security*. Washington, D.C.

Government of Canada. (2000). Schedule 1 (Section 5) Principles Set out in the National Standard of Canada Entitled Model Code for the Protection of Personal Information. *Personal Information Protection and Electronic Act (PIPEDA)*, p. 20. Retrieved from http://laws-lois.justice.gc.ca/eng/acts/P-8.6/page-20.html

Health Canada. (2011). *Toys Regulations*. Government of Canada. Retrieved from http://laws-lois.justice.gc.ca/eng/regulations/SOR-2011-17/index.html

Health Canada. (2012). *Industry Guide to Health Canada's Safety Requirements for Children's Toys and Related Products*. Health Canada. Retrieved from http://www.hc-sc.gc.ca/cps-spc/pubs/indust/toys-jouets/index-eng.php#f1

Hernan, S., Lambert, S., Ostwald, T., & Shostack, A. (2006). Uncover Security Design Flaws Using the STRIDE Approach. MSDN Magazine.

International Telecommunication Union (ITU), United Nations Children's Fund (UNICEF). (2014). *Guidelines for Industry on Child Online Protection*. Geneva, Switzerland: International Telecommunication Union. Retrieved from http://www.itu.int/en/cop/Documents/bD_Broch_INDUSTRY_E.PDF

Livingstone, S., Haddon, L., Gorzig, A., & Olafsson, K. (2011). *Risks and Safety on the Internet: The Perspective of European Children, Full findings and policy implications from the EU Kids Online survey of 9-16 year olds and their parents in 25 countries*. London: London School of Economics and Political Science. Retrieved from http://www.lse.ac.uk/media%40lse/research/EUKidsOnline/EU%20Kids%20II%20(2009-11)/EUKidsOnlineIIReports/D4FullFindings.pdf

Madden, M., Cortesi, S., Gasser, U., Lenhart, A., & Duggan, M. (2012). *Parents, Teens, and Online Privacy*. Washington D.C.: Pew Research Center, Berkman Center for Internet & Society at Harvard University. Retrieved from http://www.pewinternet.org/files/old-media//Files/Reports/2012/PIP_ParentsTeensAndPrivacy.pdf

Meier, J., Mackman, A., Dunner, M., Vasireddy, S., Escamilla, R., & Murukan, A. (2003). Chapter 3: Threat Modeling. In *Improving Web Application Security: Threats and Countermeasures*. Microsoft Corporation. Retrieved from https://msdn.microsoft.com/en-us/library/ff648644.aspx

MMA Privacy & Advocacy Committee. (2011). *Mobile Application Privacy Policy Framework*. New York, London, Singapore, Sao Paulo: Mobile Marketing Association. Retrieved from file:///C:/Users/Laura/Downloads/MMA_Mobile_Application_Privacy_Policy_15Dec2011PC_Update_FINAL.pdf

OASIS. (2013). *eXtensible Access Control Markup Language (XACML) Version 3.0*. OASIS. Retrieved January 2015, from http://docs.oasis-open.org/xacml/3.0/xacml-3.0-core-spec-os-en. pdf

Office of the Privacy Commissioner of Canada. (2012, June). *Policy Position on Online Behavioural Advertising*. Retrieved November 2014, from https://www.priv.gc.ca/information/ guide/2012/bg_ba_1206_e.asp

Office of the Privacy Commissioner of Canada. (2014, May). *Guidelines for Online Consent*. Retrieved February 2015, from https://www.priv.gc.ca/information/guide/2014/gl_oc_201405_e. asp

Open Web Application Security Project (OWASP). (2013, February). *OWASP Mobile Security Project—Mobile Threat Model*. Retrieved February 2015, from https://www.owasp.org/index. php/Projects/OWASP_Mobile_Security_Project_-_Mobile_Threat_Model

Organization for Economic Cooperation and Development. (2013). The OECD Privacy Framework. Retrieved from http://www.oecd.org/sti/ieconomy/oecd_privacy_framework.pdf

OWASP. (2013). *Application Threat Modeling*. OWASP. Retrieved from https://www.owasp.org/ index.php/Application_Threat_Modeling

Peng, H., Gates, C., Sarma, B., Li, N., Qi, Y., & Potharaju, R. (2012). Using Probabilistic Generative Models for Ranking Risks of Android Apps. *ACM Conference on Computer and Communications Security*. New York: ACM.

Pfitzmann, A., & Hansen, M. (2010). *A terminology for talking about privacy by data minimization: Anonymity, unlinkability, undetectability, unobservability, pseudonymity, and identity management (v0.34 August 2010)*. TU Dresden and ULD Kiel. Retrieved from http://dud.inf. tu-dresden.de/Anon_Terminology.shtml

Reay, I., Dick, S., & Miller, J. (2009). A Large-Scale Empirical Study of P3P Privacy Policies: Stated Actions vs. Legal Obligations. *ACM Transactions on The Web, 3*(2), 6:1–6:34.

Roosa, S. (2012, September 10). *Privacy Threat Model for Mobile*. Retrieved February 2015, from Freedom to Tinker: https://freedom-to-tinker.com/blog/sroosa/privacy-threat-model-for-mobile/

Salomon, D. (2010). Privacy and Trust. In *Elements of Computer Security, Undergraduate Topics in Computer Science* (pp. 273–290). Springer. Retrieved from http://link.springer.com/chapter/10.1007/978-0-85729-006-9_11

Schrock, A., & Boyd, D. (n.d.). *Online Threats to Youth: Solicitation, Harassment, and Problematic Content*. Cambridge: Berkman Center for Internet & Society, Harvard University. Retrieved from http://cyber.law.harvard.edu/sites/cyber.law.harvard.edu/files/RAB_Lit_Review_121808_0.pdf

Toy Industry Association. (2012). *The Changing Privacy and Data Security Landscape—From Mobile Apps to OBA*. Washington, D.C.: Keller and Heckman LLP. Retrieved from http://www.toyassociation.org/App_Themes/tia/pdfs/priorities/M2C/PrivacyWhitePaper.pdf

U.S. Department of Justice, National Sex Offender Public Website. (n.d.). *Raising awareness about sexual abuse: Facts, myths and statistics*. Retrieved March 2015, from http://www.nsopw.gov/en/Education/FactsMythsStatistics

United Nations Children's Fund (UNICEF). (1989). *United Nations Convention on the Rights of the Child*. Geneva: United Nations. Retrieved from http://www.ohchr.org/EN/ProfessionalInterest/Pages/CRC.aspx

United Nations Children's Fund (UNICEF). (2011). *Child Safety Online: Global Challenges and Strategies*. Florence, Italy: UNICEF. Retrieved from http://www.unicef.org/pacificislands/ ict_eng.pdf

United States Federal Trade Commission. (1998). *Children's Online Privacy Protection Act of 1998*. Retrieved February 2015, from http://www.coppa.org/coppa.htm

Waters, G., Wheeler, J., Westerinen, A., Rafalow, L., & Moore, R. (1999). Policy Framework Architecture. IETF. Retrieved from http://tools.ietf.org/html/draft-ietf-policy-arch-00

Westin, A. F. (1967). *Privacy and Freedom*. New York: Athenum.

Case Study: Approaching the Learning of Kanji Through Augmented Toys in Japan

Kamen Kanev, Itaru Oido, Patrick C. K. Hung, Bill Kapralos
and Michael Jenkin

Abstract Aside from their use in recreation, toys and toy technologies can also be employed in enhanced learning and education. The merging of augmented reality with traditional toys can lead, for example, to unique and engaging educational experiences providing opportunities for focused learning and more advanced knowledge dissemination. In this respect, the educational perspectives of various toys and toy technologies are considered in this work, and related instructional features and learning functionalities are presented and discussed. A research initiative led by the authors that integrates traditional toys with novel augmented reality technologies to support the learning of kanji characters, a time consuming and often difficult task, is reported and discussed.

Keywords Augmented reality (AR) toys · Computing management · Knowledge management and modeling · Mobile software development · Toy industry marketing

Introduction

According to Oxford Dictionaries, a "toy" is defined as a product or a device that is intended for use in various settings including education, training, and amusement. The economical factors of toys and toy-based technologies generate a substantive international market for toy industries (Toy Markets in the World 2013). Refer-

B. Kapralos (✉) · K. Kanev · I. Oido · M. Jenkin
Graduate School of Informatics, Shizuoka University, Hamamatsu, Japan
e-mail: bill.kapralos@uoit.edu

K. Kanev · P. C. K. Hung · B. Kapralos
Faculty of Business and Information Technology, University of Ontario Institute of Technology,
Oshawa, Canada

M. Jenkin
Lassonde School of Engineering, York University, Toronto, Canada

© Springer International Publishing Switzerland 2015
P. C. K. Hung (ed.), *Mobile Services for Toy Computing*, International Series on
Computer Entertainment and Media Technology, DOI 10.1007/978-3-319-21323-1_9

175

ring to the *Toy Industry Association* of the United States, the sales volume of toys worldwide was estimated at $ 78.1 billion (all values in US dollars) in 2007, growing by 7% over 5 years to reach $ 84.1 billion in 2012 (Toy Industry Association, Inc 2014). The world's leading toy manufacturers include Hasbro, Mattel, JAKKS Pacific, LEGO, and Namco Bandai, and their combined revenue amounted to approximately $ 20 billion in 2011. In 2010, an average of $ 284 and 317 was spent on toys per child in the Canada and the United States, respectively. The top ten worldwide markets with respect to retail toy sales are the United States, Japan, China, the United Kingdom, France, Germany, Brazil, India, Australia, and Canada (Statistics and Facts on the Toy Industry, Statista 2014).

The growing market demand for toys together with the great technological advances we have already experienced, particularly with respect to computing power and miniaturization of technology, and the opportunities these advances have created (e.g., the support of augmented reality on our smartphones), provides unique opportunities for the toy industry. Based on reporting from the *110th Annual American International Toy Fair*, in New York, United States, toy manufacturers are increasingly incorporating augmented reality (AR) technologies into their products. AR supplements reality rather than completely replacing it and can be used to enhance a user's perception of and interaction with the real world (Azuma 1997).

Augmented reality can be used to create the illusion that virtual, computer-generated objects exist in the real world and this offers an endless number of possibilities for game and toy design (Cawood and Fiala 2007). With the ongoing advances that we are experiencing in consumer adoption and interest in devices such as the Microsoft Kinect, and Google Glass, amongst others, augmented reality has become a mainstream technology. It has made its way into a number of entertainment-based applications resulting in unique and engaging interactive applications; there are many AR systems available in a toy store near you. The range of existing toys utilizing AR is substantive. AR-based games such as the 3DS Augmented Reality mini-games display 3D content through AR character cards for use on the Nintendo 3DS (Hands-on with the 3DS Augmented Reality mini-games 2014). Disney Dream Play, an interactive music toy, employs AR technologies providing the users with an enhanced experience as they interact with various Disney characters (Robertsonon 2014). In Sony PS3's Skylanders, plastic figurines are augmented with writable radio-frequency identification (RFID) cards are integrated with a video game (Tyni et al. 2013). Mattel's Disney Princess Ultimate Dream Castle interacts with an iOS application using visual tags embedded in a dollhouse. Popar Toys' have also developed a range of AR books using tablet devices.

Despite the growing popularity of augmented reality for entertainment purposes, its integration with "traditional" toys for enhanced learning has been limited (Strauss 2014; Strauss 2015). The merging of augmented reality with traditional toys can lead to unique and interactive educational applications that are able to better motivate and engage the learner. Within an educational setting, higher levels of engagement have been linked to higher academic achievement (Shute 2009). The opportunity to increase student engagement and motivation within an educational setting is the primary reason for the current interest in the use of game-based technologies (e.g., serious games) in the classroom. Game-based learning with serious games, AR toys, and other interactive media can provide students with a learner-centered approach to learning, enabling them to study via an active, critical learning approach in contrast to traditional teacher-centered approaches where the instructor dictates the learning. Student-centered approaches to learning have shown to be more effective particularly with the current generation of learners (Stapleton 2004). In addition to the direct educational benefits, the growing interest and use of such interactive media within an educational setting provides great commercial opportunities (Hinske et al. 2009).

Given the maturity of AR technology in addition to the desirability of toys and similar physical devices to enhance learning, we have begun a research initiative that will see the integration of traditional toys with novel augmented reality technologies. We are currently focusing on the application of AR toys to learning the basic symbols that comprise the writing system in kanji-related (Chinese character) languages in Japan. As described in the following sections, this can be a difficult task requiring significant time and effort. Here, we provide an overview of our on-going work, and outline issues that must be resolved before AR-based toys can be employed within an educational setting on a large scale. The remainder of this chapter is organized as follows. In Section "AR Toys and Enhanced Learning of Kanji Characters", a description of AR-based educational toys, with a focus on those intended for kanji learning is provided. A discussion regarding AR-based technologies is provided in Section "AR Enriched Viewing and Interactions". A description of an AR-based (toy) system that was developed specifically for kanji learning is provided in Section "AR Learning Experiences and AR Toys". Details regarding a preliminary experiment that was conducted to gauge user interest in the system are also provided. Section "Some Concerns and Problems to Overcome" outlines several issues that must be overcome before the use of AR-based toys within an educational setting becomes more widespread and finally, concluding remarks are provided in the final Section.

AR Toys and Enhanced Learning of Kanji Characters

With the growing numbers of mobile technology users worldwide, toy companies are employing AR technologies to tap into these new markets. However, there are many interaction and computing challenges with respect to AR toys that must be overcome before educational AR toys become more prevalent (Toys and Games 2014; Industry Guide to Health Canada's Safety Requirements for Children's Toys and Related Products 2012). More specifically, the toy industry is confronted with the challenge of requiring a better understanding of the AR consumer needs and building business models that will successfully market the AR toys. Toy makers usually require 18–24 months to develop a new product thereby implying that the development process of AR toys may be much more costly, lengthy, and difficult due to the technological challenges. Given the speed of change in portable devices including smartphones, gaming consoles, and tablets, it is difficult for game designers to plan for and cope with performance, resolution, and software infrastructure advancements.

Another exacerbating issue is that a computing management model with technological standards for educational AR toys still needs to be established within the toy industry worldwide. Furthermore, educational game industry standards must account for cultural differences and differences in educational backgrounds across many countries and regions. The authors of this paper have extensive experience both in Western and Asian cultures and are well aware of the conceptual gaps that exist between the two cultural regions. While the toy and game industries across the regions are diverse, educational similarities exist even between countries with widely different writing systems, such as the hieroglyphic writing systems of China and Japan and the Latine-based alphabet systems of countries with western European roots. Although there are of course many similarities between various writing systems, there are also marked differences.

As a concrete example, consider the problem of learning the basic symbols that make up the writing system in a kanji-related (Chinese characters) language versus the problem of learning the symbols in the Latin alphabet (English characters). To start with, the mere number of different kanji that children in Japan and China have to learn/remember is in the thousands as opposed to the 26 characters present in the English alphabet used in North America. Although a reduced Japanese alphabet of 48 hiragana characters exists, learning/remembering the 48 hiragana does not imply one is literate. Learning and mastering a large number of kanji is therefore required in certain Asian countries and a significant amount of time and effort is dedicated specifically to this task in school programs. Since the importance and the proportion of kanji learning in the context of other subjects in such countries is quite high, we consider it to be a good candidate for enhancements and optimizations with novel methods and advanced AR technologies. Given the complexity of learning the kanji character set, a range of different kanji-related (Chinese Characters) toys, educational games, and other assistive systems have been developed (Kanev and Kimura 2011; Barneva et al. 2009; Kanev et al. 2012).

In contrast to individual alphabet letters in the Latin-based alphabet, kanji characters can be associated with different objects according to their meaning and this can be used for learning enhancements in educational games. Trivial examples such as the kanji character for 'tree' (木) and its resemblance to a simplified sketch of a real tree is often provided to new learners of the language but unfortunately this simple mapping cannot be extended to more advanced levels. Modern kanji characters are often so evolved that a direct matching to what they represent is highly superficial. For example, the appearance of the kanji symbol for country (国) and its components, a mouth (口) and a ball (玉), have no apparent relation to the notion of a country. An older version of this symbol however (國) is historically described as an area (the old meaning of the inner part) surrounded by borders (the meaning of the enclosing box.) That being said, when learning kanji characters it can be useful to utilize the resemblance between the symbol and the object it represents, albeit abstractly. Given the sometimes abstract nature of this representation, it can be more instructive if a transformation process is constructed for the purpose of matching.

 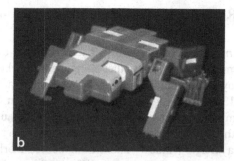

a   b

Fig. 1 A sculptured "frog" (蛙) kanji character (**a**), and a 3D frog model built by rearrangement of its parts (**b**)

To illustrate this, the example of the Mojibakeru game developed by Bandai (Mojibakeru kanji-animal transformers 2014) is used in this work.

In Fig. 1a, a sculptured kanji character denoting a frog is shown. Note that although as written characters kanji are flat (2D), the sculptured kanji shown in the figure is actually a 2.5D model that can be rearranged to appear as a sculptured 3D frog as shown in Fig. 1b. This is a good illustration of how *an abstraction* transforms into *a thing* in a tangible, physical way: An abstract kanji character that only literate people can read transforms into a more or less realistic model that everybody can recognize. In this case, an abstraction is illustrated by a physical transformation into something that is easier to understand. The disadvantage of such physical transformations is that they can be difficult to implement within an educational tool or game, and the development and production cost of the corresponding physical toy models could be prohibitive. Augmented reality and AR toys offer virtually unlimited possibilities for transformations in a cost-effective and convincing, although less tangible way. However, AR offers additional possibilities as well. With AR, an abstraction might be better illustrated by the process itself, rather than by its result, for example, in the case of kanji characters denoting verbs. Furthermore, there are many other possible representations at different abstraction levels that can be used for illustrations and providing additional views on the subject matter.

In this context, augmented reality provides yet another layer of representation; a rather promising one due to its potential to change the way we interact with the surrounding world. It offers new modalities for existing methods and technologies to reach to our senses and engage them in more effective learning. The early ideas regarding multiple view explanations (Hirotomi et al. 2003) are in line with the above given examples illustrating the meaning of kanji. With the recent technological advancements in both the virtual reality (VR) and augmented reality domains, we now have an opportunity to seamlessly blend the many different representations, views, and explanations into a unified environment where everything is (as if) at the touch of our fingers. Such an approach essentially constitutes an educational communication environment with self-explanatory components (Mirenkov et al. 2001) but in an AR-based form.

AR Enriched Viewing and Interactions

Currently, the majority of AR toys augment a camera view of the real world with some VR objects and other components. Although, it is interesting to observe that Sony's SkyLander's technology works in exactly the opposite direction (Tyni et al. 2013). That being said, consider the problem of augmenting the visual world with computer generated/presented content. The augmented view is typically presented to the user on the screen of a smartphone or a tablet (Robertsonon 2014) that serves as a "magic window" into the augmented world. The "magic" here is that what the user views on the screen is different from what is actually out there, yet more or less synchronized with the real world behind the tablet. Generating a true "magic window" involves registering the pose of the world—or at least parts of the world—with the magic window that is being viewed through, so that reality and its augmentation can be combined. Of course there are a number of different ways in which this can be operationalized, but the key technical requirement is that the display device must be able to register itself with respect to the external visual world and to then overlay content on the captured world view.

A common mechanism for a magic window display registration is to utilize a rear-facing camera mounted on the device to capture fiducial markers in the field of view. Such markers may be explicit (e.g., ARTag (Fiala 2004) or ARToolkit (Kato and Billinghurst 1999)), but it is also possible to utilize implicit markers by taking advantage of "naturally" occurring visual texture. For example, if a planar visual surface is involved and if features such as Scale-Invariant Features Transform (SIFT) (Lowe 2004) parameters can be extracted, then tracking and matching methods such as the Parallel Tracking And Mapping (PTAM) methods (Klein and Murray 2007; Castle et al. 2008), amongst others can be used to localize the feature in the image. Although this enables visual information to be overlaid onto the external world, this on its own does not make a "magic window."

In order to accomplish this, the relative pose between the viewer's eye position and the mirror must also be determined. This issue can be addressed by employing a two-way tracking approach where both a back-facing and a front-facing camera are used simultaneously on the display device. While the back camera tracks the tablet position and orientation with respect to the surrounding environment, the front camera tracks the eyes of the user. As with the problem of registering the tablet surface to the scene, the task of registering the tablet surface to the user can be accomplished in a number of ways. Although the simplest way to achieve this tracking may be to use some type of a marker attached to a pair of glasses or just glasses with a distinctive, easy to identify frame, general face tracking methods can be employed here as well. Note that in the general case, the face tracking camera and the tracked human face move differently with respect to the environment. However, since both the front and the back camera are rigidly attached to the tablet, the tablet motion determined based on the rear camera feedback can be used to transform the front camera view before the face recognition step. With this, we have a "magic window"

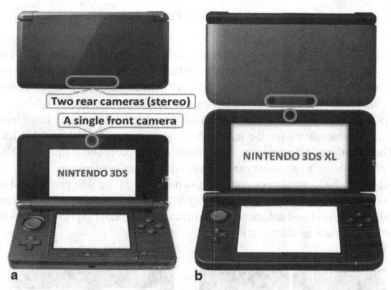

Fig. 2 Nintendo 3DS (**a**), and 3DS XL (**b**) models

that can better adhere to the physical laws of the tangible world. Through it one can see an augmented scene from different perspectives by moving the tablet around it and by looking at the screen from different angles and distances. It is important to note, however, that constructing a perfect magic mirror for multiple users will require that different views be rendered for different observers, an even more challenging problem.

As described above, the magic window is monocular; it does not properly present stereo disparity cues to a user. Although AR researchers have designed and implemented a number of stereo AR systems, developing a stereo system for the commercial education market introduces further complexity (Kato and Billinghurst 1999; Klein and Murray 2007). That being said, it is instructive to observe that a number of game platforms already exist that provide at least limited stereo viewing and capturing possibilities. Consider the Nintendo 3DS/XL models shown in Fig. 2 for example. These game platforms feature two rear cameras suitable for capturing a stereo video input of the scene, provide a single front camera for the face tracking, and come equipped with a 3D presentation screen that requires no special 3D glasses (Hands-on with the 3DS Augmented Reality mini-games 2014).

The "magic window" that we have discussed so far is unidirectional since one can only see through it from the screen side. A bidirectional magic window could be constructed from two displays placed "back to back". Such a "two way" AR device has unique properties that make it indispensable for AR interaction studies. A bidirectional "magic window" could be employed in other different modalities, including semitransparent and reflective modes. For example, it could be set up as an AR mirror on one side and an AR see-through on the other.

AR Learning Experiences and AR Toys

We have developed a number of different AR systems targeted to the task of learning kanji characters. Each of these systems differ in a variety of ways, but the underlying goal of each system is to explore how AR and physical toys can be combined to develop an effective training system for the kanji character set.

In (Kanev et al. 2012) we describe an implemented AR assistive kanji learning environment that allows for the interactive construction of AR kanji. AR kanji is based on traditional AR components that enable an appropriately equipped AR system with the ability to augment card targets with virtual content.

Designs of the kanji components as shown in Fig. 3a are first printed and registered as AR Toolkit markers. When displayed to the AR system, such markers are instantly recognized and enhanced with corresponding 2.5D representations in the respective AR view as shown in Fig. 4a.

Fig. 3 Kanji components designed as AR Toolkit markers (**a**), and a *black* and *white* picture of similar kanji cards from the popular Dr. Kanji game (**b**)

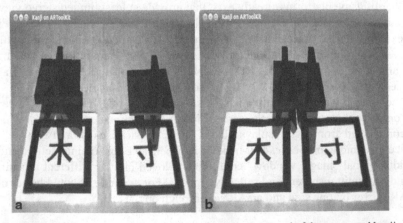

Fig. 4 AR views of the printed kanji components from Fig. 3 (**a**), and of the compound kanji automatically constructed when the components are brought together in the *right* order (**b**)

AR Kanji utilizes augmented reality to link the kanji symbol with its representative real-world symbol, where appropriate, allowing learners to explore the ways in which basic kanji symbols can be combined. For example, if compatible printed kanji components are brought together in the correct order, as shown in Fig. 4b, the corresponding compound kanji is automatically constructed and shown in the AR view.

Preliminary Experiment

The integration of AR and printed kanji symbols provides for a novel mechanism of relating the intent and/or function of a kanji character with its appearance on a simple printed card, but it is possible to imagine systems based on more complex toys where the toys themselves help to encode possible symbol combinations.

In order to explore this concept, a series of kanji learning experiments have been conducted with plastic edge-encoded kanji building cards generated using a 3D printer as shown in Fig. 5 where valid kanji constructions are confirmed by physically matching card codes. The initial system was based on planar targets where the keyed edges of the targets encoded potential symbol combinations.

A revised version of this approach utilizes 3D printing to build more sophisticated cards with 2.5D (Fig. 6) and true 3D (Fig. 7) kanji components. The enhanced cards feature multilayered matching edges with 2D codes which allow for encodings that support larger kanji construction sets.

Preliminary kanji learning experiments employing the above kanji contraction sets as kanji puzzles (Fig. 8) were conducted during an Open Campus event at Shizuoka University (Hamamatsu, Japan), November 9–10, 2013. The purpose of these experiments was twofold: i) to gauge user interest in the system, and ii) to obtain feedback regarding our system in order to assist us with further development.

The learning activity attracted 97 walk-in participants (55 on the first day and 42 on the second day). Most of the participants were children accompanied by their parents or other adults. Participants were provided with an overview of the sys-

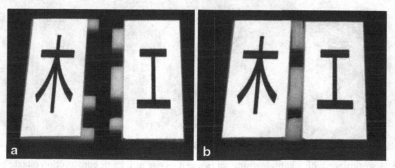

Fig. 5 Plastic edge-encoded kanji building cards printed by a 3D printer (**a**), and a valid kanji construction confirmed by the matching 1D edge codes on the cards (**b**)

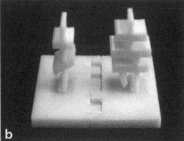

Fig. 6 Plastic edge-encoded kanji building cards printed by a 3D printer (**a**), and a valid kanji construction confirmed by their matching 2D edge codes (**b**)

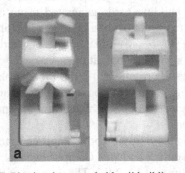

Fig. 7 Plastic edge-encoded kanji building cards printed by a 3D printer (**a**), and an invalid kanji construction confirmed by their mismatching 2D edge codes (**b**)

Fig. 8 A workplace with a kanji learning set (**a**), and its use as a kanji construction puzzle by two elementary school children (**b**)

tem and a demonstration of how to use it. They were then allowed ample time (10–20 min initial time slot which was extended if participants wanted to continue and no other visitors were waiting) to use the system either individually or in small groups for exploratory kanji learning. After this exploratory period, each participant was asked to complete a non-obligatory "fulfillment" questionnaire.

The questionnaire included four questions and was specifically designed to make a distinction between traditional learning of kanji denoting the standardized educational process in Japanese schools and other non-standardized game-based learning activities such as our kanji construction puzzle game and the AR kanji game. Question 1 (Q1) aimed to determine whether the participant had been previously exposed to the standard school-bound kanji learning process and if so, their level of satisfaction. Q2 and Q3 shifted the focus and set apart play and study activities as perceived by the participants. In this regard, Q2 focused on recreational play that may help in learning kanji while Q3 focused on regular kanji learning by playing a game. Finally, Q4 prompted for feedback on the participants' expectations and game activity preferences based on their experience with our experimental kanji games.

Results

Approximately 42% of the respondents were elementary school students (lower grades- 12%, higher grades- 30%) with limited kanji knowledge. The other participants were mostly secondary and high school students and some adult non-Japanese visitors that were learning kanji. Participant responses from completed questionnaires are summarized in Table 1.

Table 1 Summarized responses to the fulfilment questionnaire completed after the learning sessions with the kanji construction sets

Q1. How do you find the traditional learning of kanji?	
A). Boring	18%
B). Neutral	59%
C). Exciting	23%
Q2. How do you like the kanji construction puzzle game?	
A). Not interested	6%
B). Neutral	0%
C). Interested	94%
Q3. Would you like to use the kanji construction puzzle game for regular kanji learning?	
A). No	0%
B). From time to time	23%
C). Yes, always	77%
Q4. Are you going to learn kanji better if games are used?	
A). Yes, with the kanji construction puzzle game	73%
B). Yes, with the AR kanji game	20%
C). No	7%

The majority of the respondents were neutral with respect to traditional methods of learning of kanji (Q1B, 59%) while the rest were almost equally divided between those who expressed a negative attitude (Q1A, 18% found it boring) and a positive attitude (Q1C, 23% found it exciting). While most of the participants enjoyed the kanji construction set as a game and found it interesting (Q2C, 94% interested vs. Q2A, only 6% not interested), a smaller percentage of them (Q3C, 77%) definitely wanted to use our system for regular kanji learning. Nevertheless, 93% (Q4A and Q4B) of the respondents believed that they could learn better if games were incorporated into the kanji learning process (Q4, 73% were in favor of the kanji construction puzzle game and 20% were in favor of the AR kanji game).

Discussion

Production of the edge-encoded kanji building cards in Figs. 5, 6 and 7 require 3D printing and substantial investment in design and materials while conventional printing on regular paper is sufficient for the AR Toolkit markers in Fig. 4 which helps to demonstrate some of the advantages of the AR approach. The kanji printouts are being adapted to AR Toolkit traceable markers and employed in AR training. However, 3D kanji constructs expose different, direction dependent views and this complicates AR recognition and tracking. In order to address this issue, we are investigating and experimenting with alternative AR target tracking including surface encoding, SIFT, PTAM, and other related approaches.

In an ideal world, a user should be able to play with different 3D kanji components in an AR enhanced kanji construction environment in a natural unrestricted way. During play, timely context dependent advice and pedagogically correct guidance should be provided by the AR system. To address such issues, identification, recognition, and tracking of 3D kanji components must be conducted in real-time. Our initial experiments have indicated that plastic 3D kanji printed by a 3D printer, such as those shown in Fig. 9, appear too featureless for direct handling by the SIFT and PTAM algorithms.

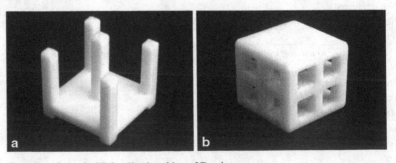

Fig. 9 Samples of plastic 3D kanji printed by a 3D printer

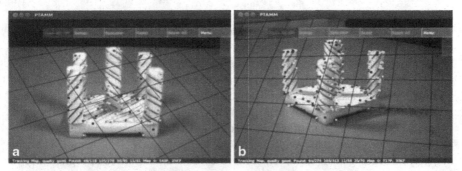

Fig. 10 A 3D kanji registration with PTAM (**a**), and its 3D reconstruction (**b**)

Experimental patching of their surfaces with manually applied texture patterns leads to a significant improvement in object registration (Fig. 10a) while recognition and tracking need furthered enhancements. Surface patched 3D kanji components have also been subjected to 3D reconstructions for identification and tracking as shown in Fig. 10b.

We are currently investigating approaches that integrate the SIFT/PTAM and 3D reconstruction approaches for more reliable component recognition and consecutive tracking. Another research direction involves the employment of kanji constructs with digitally encoded semantic surfaces (Kanev and Mirenkov 2010). Such digital encoding could serve as specialized surface patching for SIFT/PTAM and 3D reconstruction. The special properties of digital encoding include: i) it is practically undetectable by the naked eye so construct appearance to humans remains unaltered, and ii) it can carry surface related position and orientation information for direct object recognition and tracking (Gnatyuk et al. 2012; Kanev 2012).

Some Concerns and Problems to Overcome

AR-enabled toys provide a powerful mechanism to relate physical objects (toys) with complex concepts. The problem of learning kanji provides a suitable task to illustrate how physically embodied objects representing individual symbols can be combined with AR-presented concepts.

We have described some of our efforts in this regard, working from simple 2D card-displays of kanji characters coupled with traditional AR to complex 3D symbols that present more challenging AR tasks. The work is ongoing with the basic premise—physically embodied toys can be enabled with augmented-reality to enhance the ability for students to learn complex tasks. Preliminary results indicate that our AR-based kanji learning system is effective and well received by users. However, the adoption of technologies such as AR/VR and video games in the classroom is not without complications. More specifically, the use of (entertainment-based) video games have been correlated to decreased academic performance, decreased prosocial behavior, aggressive thoughts, feelings, and behaviors, smoking, and obesity (Gentile et al. 2004; Anderson and Gentile 2008). This has led some to question the benefit of video game play in general, and to automatically reject the application of video game-based technologies in the classroom.

In the development and deployment of technologically enhanced devices for training in the classroom it is critical to understand how different cultures and governments treat such deployments. Globally, there is an extensive list of banned video games and some countries have specific reasons for these bans (i.e., Germany has banned the video game Wolfenstein due to the game's Nazi theme). The vast majority of video games, however, seem to be banned in particular regions due to their excess of "blood and gore", violence, or pornographic images. The makers of AR toys must be particularly careful while attempting to enter markets with strict anti-shooting game regulations (e.g., in November 2009, the government of Venezuela banned all video games whose objective was to shoot people).

Technologically, augmented ("serious") toys can capture substantive amounts of information about the students that are using them. Some AR toys are physical embodiments that act as a user interface for information services. An AR toy can capture a user's physical activity state (e.g., walking, standing, running, etc.), and store personalized information (e.g., location, activity pattern, etc.). For example, TekRecon is an AR toy in the form of its two blasters—the pistol-looking Hammerhead and the rifle-like Havok integrated with its iOS and Android app as shown in Fig. 11. The app uses both mobile and GPS technology to augment play with the AR toy (TekRecon Advanced Battle Systems 2014).

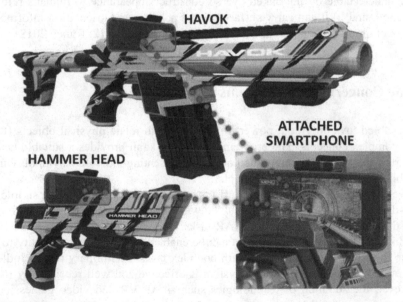

Fig. 11 TekRecon AR Toys and an attached smartphone

Toy companies are confronted with the challenge of better understanding consumer and government concerns, while at the same time exploring the possibility of adopting such context-awareness through wearable AR toys with information interfaces to other services, such as social networking and cloud computing (SOA4D Forge 2014). On one hand, some information must be kept private from others while on the other hand, some information has to be released for acquainting information or services. Companies or service providers with a good reputation for privacy protection will find it easier to develop a trust relationship with their customers. Developers must also be cognizant that the intended user of these toys are typically children and adolescents and may not be aware of the privacy implications of their interactions with a device. Furthermore, they are most likely unable to legally waive their privacy rights themselves.

A further challenge relates to privacy. Privacy can be described by the ability to have control over the collection, storage, access, communication, manipulation, and disposition of data (Hung and Cheng 2009a). Some refer to privacy as the right for individuals to determine for themselves, when, how, and to what extent information about them is communicated to others (Hung et al. 2009b). In general, a privacy policy sets out ground rules for how private sector organizations may collect, use, or disclose personal information in the course of commercial activities. Failing to comply with these legislations, in their respective countries, may lead to civil and/or criminal penalties. In addition to the penalties, organizations may suffer the loss of reputation and goodwill when the non-compliance of legislation is publicized. An exacerbating issue is that a privacy policy framework with technological standards for protecting user's location-based information on mobile services for AR toys has not yet been established within the toy industry worldwide.

Mobile services architectures are typically built on an insecure, unmonitored and shared environment, which is open to events such as security threats. As is common in many other applications, the location-based information processed in mobile services might be sensitive so it is important to protect it from security threats such as disclosure to unauthorized parties, especially for protecting children. Within the scope of this research, we are designing a privacy policy framework for protecting location-based information on mobile services during interactions with online services for AR toys. In this framework, a privacy policy expresses what the data protection mechanisms are to achieve through security measures. The framework must provide the user with the complete personalization and control facilities. Users can configure the functionalities and notification mechanisms according to their own preferences, and such configuration must obtain positive authorization from a user who is able to make the decision legally.

Conclusion

Globally, the toy industry is a multi-billion dollar per year industry and it is continuously growing. The Recent technological advancements have seen the proliferation of consumer level devices such as the Microsoft Kinect, and Google Glass, bringing augmented reality (AR) to the mainstream. AR has been applied in a large number of entertainment-based applications resulting in unique and engaging interactive experiences for the user. Given the worldwide demand for toys and the growing popularity of augmented reality, in addition to the use of interactive media such as serious games and virtual simulations in the classroom, we believe there is great opportunity for the merging of toys with augmented reality technologies to provide a richer, interactive, engaging, and motivating educational environment.

To this end, in this work we have described our ongoing research that is investigating such merging of toys and augmented reality for educational purposes with the goal of providing learners with a more engaging and motivating learning environment that allows for the simple creation of a variety of learning scenarios in a cost-effective manner. We have highlighted some of our own work that has examined augmented reality-based toys within a kanji learning environment and our preliminary results are promising. Finally, despite the potential benefits of augmented reality-enhanced toys, one should also be aware that there are also some problems that must be overcome before AR-based toys become widespread. Some of these problems are technical while others are related to social and legalistic frameworks.

Acknowledgement This work was partially supported by KAKENHI Grant Number 25560109. Part of the research is based on the Cooperative Research Project of the Research Institute of Electronics, Shizuoka University. The financial support of the Natural Sciences and Engineering Research Council of Canada (NSERC) is also acknowledged.

References

Toy Markets in the World 2013, 2013. URL: http://www.toy-icti.org/PDFs/ToyMarkets13.pdf, Retrieved June 4, 2014.

Toy Industry Association, Inc., URL: http://www.toyassociation.org/, Retrieved June 4, 2014.

Statistics and Facts on the Toy Industry, Statista, URL: http://www.statista.com/topics/1108/toy-industry/, Retrieved June 4, 2014.

Azuma, R.: A Survey of Augmented Reality, Presence: Teleoperators and Virtual Environments, 66(4):, 355–385, (1997)

Cawood, S., Fiala, M.: Augmented Reality: A Practical Guide, Pragmatic Bookshelf, (2007)

Hands-on with the 3DS Augmented Reality mini-games, URL: http://www.pocketgamer.co.uk/r/3DS/Nintendo+3DS/feature.asp?c=28176, Retrieved June 4, 2014.

Robertsonon, A.:Disney's new augmented reality toys are promising but frustrating, The Verge, URL: http://www.theverge.com/2013/1/10/3863688/disney-augmented-reality-toys-are-promising-but-frustrating, Retrieved June 4, 2014.

Tyni, H., Kultima, A., Mäyrä, F.: Dimensions of Hybrid in Playful Products, In: *Proc. International Conference on Making Sense of Converging Media* (AcademicMindTrek '13), pp. 237–244, New York, NY, USA, (2013).

Strauss, M.:As children pine for electronics, traditional toy makers face a growing tech challenge, The Globe and Mail, November 24, 2012, The Globe and Mail, URL: http://www.theglobeandmail.com/report-on-business/as-children-pine-for-electronics-traditional-toy-makers-face-a-growing-tech-challenge/article5621174/, Retrieved July 15, 2015.

Strauss, M.: TOYS Tough economy leads to trying times for toy makers, The Globe and Mail, November 15, 2012, URL: http://www.theglobeandmail.com/report-on-business/tough-economy-leads-to-trying-times-for-toy-makers/article5357558/, Retrieved June 4, 2014.

Shute, V., Ventura, M., Bauer, M., Zapata-Rivera, D.: Melding the power of serious games and embedded assessment to monitor and foster learning: flow and grow, In: U. Ritterfeld, M. Cody, and P. Vorderer, (Eds.), Serious Games: Mechanisms and Effects, Routledge, Taylor and Francis, pp. 295–321, Mahwah, NJ, USA, (2009)

Stapleton, A.J.: Serious games: serious opportunities. In: Proc. 2004 Australian Game Developers' Conference, pp. 1–6. Melbourne, Australia, (2004)

Hinske, S., Lampe, M., Price, S., Yuill, N., Langheinrich, M.: Kingdom of the Knights: Evaluation of a Seamlessly Augmented Toy Environment for Playful Learning, In: Proc. 8th International Conference on Interaction Design and Children, Milano, pp. 202–205, Como, Italy, (2009)

Toys and Games, URL: http://toysandgamesmagazine.ca/, Retrieved June 4, 2014.

Industry Guide to Health Canada's Safety Requirements for Children's Toys and Related Products, 2012, Health Canada, Government of Canada. URL: http://www.hc-sc.gc.ca/cps-spc/pubs/indust/toys-jouets/index-eng.php#a31, Retrieved June 4, 2014.

Kanev, K., Kimura, S.: Collaborative Learning in Dynamic Group Environments, In: Q. Jin (Ed.), Distance Education Environments and Emerging Software Systems: new Technologies, IGI Global, pp. 1–14, (2011)

Barneva, R., Brimkov, V., Kanev, K.: Combining Ubiquitous Direction-Sensitive Digitizing with a Multimedia Electronic Dictionary for Enhanced Understanding, International Journal of Imaging Systems and Technology, 19(2): pp. 39–49, (2009)

Kanev, K., Oido, I., Yoshioka, R., Mirenkov, N.: Employment of 3D Printing for Enhanced Kanji Learning, In: Proc. The Joint International Conference on Human-Centered Computer Environments HCCE 2012, pp. 165–170, Aizu-Wakamatsu, Japan, (2012)

Mojibakeru kanji-animal transformers, http://pinktentacle.com/2010/05/mojibakeru-kanji-animal-transformers/, Retrieved June 4, 2014.

Hirotomi, T., Mirenkov, N., Fujita, K.: Filmification of Words and Sentences Towards Teaching and Learning, In: Proc. International Conference. on Advances in Infrastructure for e-Eletronic, e-Business, e-Education, e-Science, e-Medicine on Internet (SSGRR 2003w), L'Aquila, Italy, (2003)

N. Mirenkov, A. Vazhenin, R. Yoshioka, T. Ebihara, T. Hirotomi, and T. Mirenkova, Self-explanatory components: a new programming paradigm, Int. J. of Software Engineering and Knowledge Engineering, Vol. 11, No. 1, World Scientific, 2001, 5–36.

Fiala, M., ARTag Revision 1. A Fiducial Marker System Using Digital Techniques. Technical Report, National Research Council of Canada, Ottawa, Canada, (2004)

Kato, H., Billinghurst, M., Marker tracking and HMD calibration for a video-based augmented reality conferencing system, In: Proc. 2nd IEEE and ACM International Workshop on Augmented Reality (IWAR 99), San Francisco, CA, (1999)

Lowe, D.: Distinctive image features from scale-invariant keypoints, International journal. of Computer Vision, 60(2): 91–110, (2004).

Klein, G., Murray, D.: Parallel Tracking and Mapping for Small AR Workspaces, In: Proc. International Symposium on Mixed and Augmented Reality (ISMAR 2007), pp. 1–10, Nara, Japan, (2007)

Castle, R., Klein, G., Murray, D.: Video-rate Localization in Multiple Maps for Wearable Augmented Reality, In: Proc. 12th IEEE International Symposium on Wearable Computers, Pittsburgh PA, (2008)

Kanev, K., Mirenkov, N.: Pervasive Carpet Encoding for Active Knowledge Semantic Surfaces, In: Q. Li and T. K. Shih (Eds), Ubiquitous Multimedia Computing, Chapman & Hall/CRC Press, pp. 197–213, (2010)

Gnatyuk, V., Kanev, K., Mizeikis, V., Aoki, T., Gagarsky, S., Poperenko, L.: Laser Volumetric
    Marking and Recording of Digital Information, In: Proc. The 10th International Conference on
    Global Research and Education (InterAcademia 2012), pp. 189–200, (2012)
Kanev, K.:Augmented Tangible Interface Components and Image Based Interactions, In: proc.
    International Conference on Computer Systems and Technologies (CompSysTech'12), pp. 23–
    29, Ruse, Bulgaria, (2012)
Gentile D.A., Lynch P.J., Linder J.R., Walsh D.A., The effects of violent video game habits on
    adolescent hostility, aggressive behaviors, and school performance. Journal of Adolescence,
    27(1): 5–22, (2004)
Anderson, C. A., Gentile, D. A.: Media Violence, Aggression, and Public Policy. In: E. Borgida,
    and S. Fiske (Eds)., Beyond Common Sense: Psychological Science in the Courtroom, Wiley,
    (2008).
TekRecon Advanced Battle Systems. URL: http://www.tekrecon.com/, Retrieved June 4, 2014.
SOA4D Forge. URL: https://forge.soa4d.org/, Retrieved June 4, 2014.
Patrick C. K. Hung and Vivying S. Y. Cheng, "Privacy," the Encyclopedia of Database Systems
    (EDS), Pages 2136–2137, Springer, 2009a
Patrick C. K. Hung, Yi Zheng, and Stephanie Chow, "Privacy Policies and Preferences," the Ency-
    clopedia of Database Systems (EDS), Pages 2140–2142, Springer, 2009b

Printed in the United States
Dy Dookmasters